The Cuneiform Uranology Texts

Drawing the Constellations

Paul-Alain Beaulieu
Eckart Frahm
Wayne Horowitz
John Steele

Transactions of the
American Philosophical Society
Held at Philadelphia
For Promoting Useful Knowledge
Volume 107, Part 2

ISBN: 978-1-60618-072-3
US ISSN: 0065-9746

Library of Congress Cataloging-in Publication Data

Names: Beaulieu, Paul-Alain, author. | Frahm, Eckart, author. | Horowitz, Wayne, 1957- author. | Steele, John M., author.
Title: The cuneiform uranology texts : drawing the constellations / Paul-Alain Beaulieu, Eckart Frahm, Wayne Horowitz, John Steele.
Description: Philadelphia : the American Philosophical Society, [2018] | Series: Transactions series, ISSN 0065-9746 | Includes bibliographical references and index.
Identifiers: LCCN 2018015145 | ISBN 9781606180723
Subjects: LCSH: Astronomy, Assyro-Babylonian. | Astronomy, Ancient. | Cuneiform writing.
Classification: LCC QB19 .B43 2018 | DDC 523.80935–dc23 LC record available at https://lccn.loc.gov/2018015145

Also available as an ebook (ISBN: 978-1-60618-077-8)

Preface and Dedication

In 1927, the great Ernst Weidner published an article titled "Eine Beschreibung des Sternenhimmels aus Assur," in *Archiv für Orientforschung* (vol. 4, pp. 73–85). This article presented a most interesting tablet from Assur in Neo-Assyrian script describing how to draw a number of the Mesopotamian constellations including "The Great Twins" and "The Lesser Twins" (Gemini), "The Crab" (Cancer), "The Lion" (Leo), and "The Wagon" and "The Wagon of Heaven" (the Big and Little Dippers). For most of the 20th century, this remained a unique member of the cuneiform corpus. Then, in the 1990s, working independently, Paul-Alain Beaulieu identified three more exemplars of the same type of text in the tablet collections of Yale University, and Wayne Horowitz identified another exemplar at The British Museum, thus giving us a group of five tablets of the type published by Weidner. In July 2005, at the "Calendars and Years" workshop of the Seventh Biennial History of Astronomy Workshop held at Notre Dame University, the topic of Weidner's tablet came up during a discussion between Horowitz and the late John Britton, who knew of Beaulieu's work on the Yale exemplars. The result was to be a joint publication of the group by Beaulieu, Britton, and Horowitz, for which Britton was to provide the astronomical expertise and the two other authors would be responsible for tablets on which they were already working: Beaulieu, the tablets at Yale; and Horowitz, a re-edition of Weidner's tablet with the British Museum duplicate. This was not to be, due to the untimely passing of John Britton. Fortunately, John Steele agreed to take the place of John Britton as the project's astronomical expert, and preparation for publication continued apace. In recent years, Eckart Frahm of Yale University has joined the group and made important contributions regarding the tablet fragments at Yale; in particular, producing a handcopy and edition of MLC 1884 and undertaking a detailed study of the non-astronomical part of this tablet (he is solely responsible for Appendix C). This brings our team of scholars to the four named on this book's title page, but in reality we are five. The fifth is our missing friend, John Britton, to whom this volume is dedicated and without whom this volume would never have come to be. May his memory be blessed.

Conventions

Below we use the English term "star" in the same way that Ancient Mesopotamians used Sumerian mul = Akkadian *kakkabu*. These cover a much wider variety of objects in the sky than modern English "star" in typical daily usage. Sumerian mul = Akkadian *kakkabu* can be used for individual fixed stars, planets, constellations, and constellation parts, but also for all sorts of other observable phenomena in the sky including, but not only, comets, meteors, and mirages. We will refer below to individual stars, such as modern Sirius, Polaris, or Alpha Centauri, as "fixed stars."

Unless otherwise noted, Mesopotamian star-names and their translations are given as in the ancient source material, and with reference to the star list of Urra XXII published in Bloch and Horowitz (2015), the reference work of Kurtik (2007), and the Star Catalog of BPO 2 (Reiner and Pingree 1981, 9–16). Identifications of ancient names with modern stars and constellations are as in Kurtik (2007) and/or BPO 2. Standard Assyriological abbreviations are as in the CAD (*Chicago Assyrian Dictionary*) or PSD (*Pennsylvania Sumerian Dictionary*). The abbreviation *Horowitz Alb* is for Horowitz (2014), and GSL is for *The Great Star List*.

When used as an item marker, we render the DIŠ sign as ¶.

Italics are used within translations to indicate uncertainty of meaning.

Acknowledgments

We thank the Vorderasiatisches Museum, Berlin; the British Museum, London; and the Yale Babylonian Collection, New Haven for permission to study and publish the tablets that are the subject of this book. Wayne Horowitz's work on these tablets was in part supported by a Franklin Research Grant from the American Philosophical Society.

Contents

CHAPTER 1

Introduction

Our book, *The Cuneiform Uranology Texts: Drawing the Constellations*, presents a newly recovered group of cuneiform texts from first millennium Babylonia and Assyria that provide prose descriptions of the drawing (*eṣēru*) of Mesopotamian constellations. The group describes these constellations in terms of their parts: body parts for constellations in human or animal form, parts of a wagon for "The Wagon" and "The Wagon of Heaven" (the Big and Little Dipper), and so forth. The descriptions also typically speak of the clothing that constellations in human form wear, their beards if they are male, and paraphernalia that they hold or carry. In the case of "The Crab" and "The Wagon," there is also reference to the Babylonian geometric shape *apsamakku*, this being a four-sided figure.

There is every reason to believe that ancient Mesopotamian stargazers saw the constellations in the sky at night along the lines of the descriptions given in our text group. Many of the constellations were probably already defined and named by the Sumerians long before the names appeared in writing. Lists of "stars" (Sumerian mul = Akkadian *kakkabu*), a term that can be used for individual fixed stars, planets, and constellations or other objects in the sky, are attested already in the early second millennium B.C.E. From later in the second millennium, we find lists of "stars" (in fact mostly constellations) arranged into groups of three stars for each of the twelve months of the year in texts known today as *Astrolabes* or "Three Stars Each."[1] The three stars for each month in these texts are assigned to one of three "paths" across the sky named after the gods Anu, Enlil, and Ea.

A longer list of stars appears in a more extensive work which reached its final form in the early first millennium known as *Mul.Apin*.[2] This is the central text of what might be thought of as "traditional" Mesopotamian astronomy, which is largely concerned with the relative sequences and dates of visibilities of the stars; the successive culminations (*ziqpu*) of the stars; and simple mathematical schemes for the length of daylight, the visibility of the moon, and the length of the shadow cast by a gnomon. Related to *Mul.Apin* are a number of other texts dealing with stars, including lists of *ziqpu*-stars, often with a statement of the number of UŠ

[1] Editions of all known Astrolabe texts and a detailed study of the genre may be found in Horowitz (2014), designated "Horowitz Alb" throughout this book.

[2] *Mul.Apin* is edited and translated with a brief commentary in Hunger and Pingree (1989); an updated edition and detailed commentary is currently being prepared by H. Hunger and J. M. Steele. For further discussion of the series, see Hunger and Pingree (1999, 57–83) and Watson and Horowitz (2011).

"degrees" between the culmination of the stars;[3] the so-called *GU Text*,[4] a text that lists strings of stars of similar right ascension, each string headed by a *ziqpu*-star; and the poorly understood *DAL.BA.AN.NA Text*.[5]

Our group makes note of most of the best known constellations from *Mul.Apin*. In *Mul.Apin*, it is these constellations, more than the individual stars that comprise them, that stand at the heart of the astronomy. In particular, *Mul.Apin*, and also other works of "traditional" cuneiform astronomy and astrology, only rarely speak of individual stars by name. Instead, individual fixed stars are normally noted with reference to the constellations they belong to, or a constellation nearby. For example, in *Mul.Apin* we find "The Star Which Stands in the Breast of the Lion," "The Star Which Stands in the Cart-pole of the Wagon" and "The Bright Red Star Which Stands in the Kidney of the Stag" (*Mul.Apin* I i 9, 26, 36), and "The Star Which Stands in Front of the Wagon, the Ewe" (*Mul.Apin* I i 18).

In the Late Babylonian period, the traditional astronomy of *Mul.Apin* existed side by side with, for want of a better term, more "scientific" strands of astronomy, which included the precise observation of astronomical phenomena, the prediction of future astronomical phenomena through the use of planetary and lunar cycles, and the development of mathematical astronomy. In contrast to the "traditional" astronomy, this latter astronomy placed greater emphasis on individual stars rather than constellations. Nevertheless, even here, stars were generally identified as part of a constellation. For example, most of the so-called "Normal Stars" used as reference points to track the positions of the moon and planets in the Astronomical Diaries and related texts are named by reference to constellations: "The Bright Star of the Ribbon of the Fishes," "The Front Star of the Head of the Hired Man," "The Rear Star of the Head of the Hired Man," and so forth.[6] This view is consistent with how Mesopotamians understood the history of the starry sky, believing that the gods arranged the individual stars into the shape of the constellations in earliest times, the era of creation. This notion is made explicit in the prologue to the astronomical/astrological compendium *Enūma Anu Enlil*, which states that the three great gods of the traditional Sumero-Akkadian pantheon, Anu, Enlil, and Ea, drew the constellations in earliest times as something akin to portraits of themselves in the sky:

e-nu ^{d}*a-nu* ^{d}*en-líl u* ^{d}*é-a ilānu*^{meš} *rabûtu*^{meš}

šamê^{e} *u erṣeta*^{ta} *ib-nu-ú ú-ad-du-u gis-kim-ma*

ú-kin-nu na-an-za-za [*ú-š*]*ar-ši-du gi-is-ga-la*

ilāni^{meš} *mu-ši-tim ú-*[x-x]-x *ú-za-i-zu har-ra-* ⌜*ni*⌝

kakkabāni tam-ši-li-[*šu-nu uṣ-ṣ*]*i-ru lu-ma-a-*[*ši*]

mūšu ūma mal$^{?}$*-ma*$^{?}$*-*[*liš im-du-d*]*u ar-ha u šatta ib-nu-u*

---

[3] See Steele (2014) for a discussion of the *ziqpu*-star lists and a summary of the stars found in these lists.
[4] *The GU Text* is edited in Pingree and Walker (1988). See also Hunger and Pingree (1999, 90–100) and Steele (2017a, 95–6) for discussions of this text.
[5] *The DAL.BA.AN.NA Text* is edited in Walker (1995). See also Koch (1995) and Hunger and Pingree (1999, 100–11) for discussions of this text.
[6] For a detailed study of the Normal Stars and a list of the stars that make up this group, see Jones (2004). Two partially preserved lists of Normal Stars are known from the tablets BM 36609+ and BM 46083, both of which are edited and discussed in Roughton et al. (2004).

When Anu, Enlil, and Ea, the great gods,
built heaven and earth, fixed the astronomical signs,
established the stellar positions, [se]t fast the stellar locations
the gods of the night they . [. .] ., divided the (stellar) paths
the gods, the likenesses [of them they dr]ew, the constellations
night (and) day, as equa[ls$^?$ they measure]d, month and year they created.[7]

A similar account of the creation of the constellations is found in the learned religious mystical work KAR 307 from Neo-Assyrian period Assur, which states that Bel/Marduk, the Babylonian King of the Gods, drew the constellations on the face of the sky, here the lowest of three heavens:

šamû šaplûtu ašpû ša kakkab āni lumāšī ša ilāni ina muhhi ēṣir

The Lower Heavens are (made of) Jasper, they belong to the stars. He (Bel/Marduk) drew the constellations of the gods on it.[8]

This passage offers a short explication of a longer tradition known from the Babylonian national epic *Enuma Elish*, in which Marduk arranges the stars in the sky as constellations at the start of Tablet V, before drawing the boundary lines that divided the sky into the 36 sectors of the Astrolabe tradition:[9]

1.	*ú-ba-aš-šim man-za-za*	*an ilāni rabûti*
2.	*kakkabāni*^{meš} *tam-šil-šu-nu*	*lu-ma-ši uš-zi-iz*
3.	*ú-ad-di šatta* (mu-an-na)	*mi-iṣ-ra-ta ú-aṣ-ṣir*
4.	12 *arhāni*^{meš}	*kakkabāni*^{meš} ⌈*šu-lu*⌉*-šá-a uš-zi-iz*
5.	*iš-tu* u_4*-mi ša šatti* (mu-an-na)	*uṣ-ṣ[i-r]u ú-ṣu-ra-ti*
6.	*ú-šar-šid man-za-az* ^{d}*Né-bé-ri ana*	*ud-du-u rik-si-šú-un*
7.	*a-na la e-piš an-ni*	*la e-gu-ú ma-na-ma*
8.	*man-za-az* ^{d}*En-líl u* ^{d}*É-a*	*ú-kin it- ti-šú*

1. He fashioned heavenly stations for the great gods,
2. And set up constellations, the patterns of the stars.
3. He appointed the year, marked off divisions,
4. And set up three stars each for the twelve months.
5. After he had organized the year,
6. He established the heavenly station of Neberu to fix the stars' intervals.
7. That none should transgress or be slothful
8. He fixed the heavenly stations of Enlil and Ea with it.

[7] Translation adapted from Horowitz (2011, 147), with bibliography for the passage available there and on p. 413.
[8] KAR 307 33 (Horowitz 2011, 3).
[9] Transliteration and translation adapted from Lambert (2013, 98–9). For this passage from *Enuma Elish* and the Astrolabes, see Horowitz Alb (1) and previously Horowitz (2007).

Such drawings of constellations, and boundary lines between sectors of the sky, are to be found on cuneiform tablets as well as in the heavens. The surviving fragments of the Astrolabe planispheres published as CT 33 11–12 present a drawing of a round Mesopotamian sky along the lines of the Astrolabe system described in *Enuma Elish*. When complete, these planispheres consisted of a circle divided into 36 stellar sectors by concentric circles and radii, thus yielding one sector for each of the 36 stars of the Astrolabes.[10] Likewise, the planisphere CT 33 10 (= K. 8538) divides the sky into eight sectors. Within each sector are rough sketches of constellations formed by dots (representing the stars) connected by lines, but these configurations do not match anything described in our group, suggesting that CT 33 10 belongs to a different strand of tradition.[11] Another planisphere, from Sippar, published in Horowitz and Al-Rawi (2001), also divides the sky into twelve sectors by means of twelve radii, placing a drawing of a rosette in the center of the sky, with the names of constellations, and stars represented by dots, along the outer rim of the planisphere. Here, the number of dots (= stars) for each constellation usually matches the number of stars counted for constellations on the *ziqpu*-star text VAT 16436 and its newly identified duplicate BM 37373.[12]

A much more elaborate set of drawings of constellations appears on two Uruk manuscripts, VAT 7851 and VAT 7847 + AO 6448 from the series of micro-zodiac texts first edited by Weidner (1967).[13] These texts contain astrological associations for each sign of the zodiac and its division into twelve "micro-zodiac" parts, each side of a tablet being concerned with one zodiacal sign. Several fragments of this material are known, among which two tablets include drawings representing constellations, the planets, and the moon. Rather than rough sketches of the constellations formed by connecting the dots with lines as in CT 33 10, here we have elaborate pictures of the constellations, drawn from a side-view perspective and identified by cuneiform labels (plate 1). The drawings are figurative and do not include individual stars, except in the case of the constellation "The Stars" (the Pleiades), which is represented by a group of seven stars. The two tablets both come from Seleucid Uruk but were written by different scribes, and the drawings on VAT 7851 are not as well executed as those on VAT 7847 + AO 6448. The obverse of VAT 7851 concerns the zodiacal sign of Taurus and preserves drawings of (from left to right) "The Stars" (labeled mul-mul); the moon, drawn as a circle enclosing the figure of a man holding in one hand a weapon of some sort and in the other the body of a lion (no label); and "The Bull of Heaven" (no label preserved). The obverse and reverse of VAT 7847 + AO 6448 concern, respectively, the zodiacal signs Leo and Virgo. For Leo, we find on the left side an eight-pointed star labeled Jupiter (^{d}sag-me-gar) and to the right a drawing of a lion labeled "The Lion" (^{mul}UR.MAH), with a lion's mane, four legs, and a tail, on top of a horned reptile with feet and wings labeled "The Snake" (^{mul}MUŠ). For Virgo, we find the tail of the reptile with a bird-form representing "The Raven" (labeled ^{mul}UGA^{mušen}) standing on the reptile's tail,

---

[10] For a reconstruction, see Horowitz Alb (1).

[11] The most recent edition is Koch (1989, 56–113). See previously Weidner (1915, 107–12).

[12] Schaumberger (1952, 226–7); Fincke and Horowitz (forthcoming).

[13] A new edition and study of the micro-zodiac texts, including several new sources from Babylon, may be found in Monroe (2016). None of the new sources of this series, however, includes the drawings, although a few preserve labels that serve as indications of where drawings would be positioned had they been included.

an eight-pointed star labeled Mercury (^{d}gu$_4$-utu), and a female holding a sheaf of barley representing "The Furrow" (^{mul}AB.SÍN). The planets and the moon appear in the position of their *bīt niṣirti*, "secret place," a position in the sky where they possess particular astrological significance.[14]

We shall see in our following discussions that these drawings on the micro-zodiac tablets provide a very good match to what is described in our group. For example, in D Section I (D iii 1–3) "The Snake" is described as winged, with feet, and with a raven on top of its tail. Another match may be found in the description of "The Lion" in ABC Section V, which makes note of the tail of "The Lion" and *hu-ru-u*[*p-p*]*i*. The tail of "The Lion" is very pronounced in the drawing on VAT 7847 + AO 6448, and the *hu-ru-u*[*b-b*]*i* can be identified with the feature drawn on the thigh of "The Lion." Thus, we may presume that the tradition of drawing constellations, and describing such drawings in writing, goes back at least to the Neo-Assyrian Period date of our Source A. Indeed, it seems possible that the drawings found on the micro-zodiac tablets existed separately from the micro-zodiac material. Several sources for the micro-zodiac series clearly did not include the drawings, and the drawings are not strictly connected to the micro-zodiac schemes, which are constructed from the zodiac of twelve signs. The two Uruk examples should perhaps be seen as "deluxe" versions of the series in which drawings that have only a visual connection to the topic of the text have been added. In particular, the drawings refer to lists of constellations which pre-date the zodiac, whose order (and particularly the placement of the planets in their *bīt niṣirti* between the constellations) is already given in our Neo-Assyrian Source A and in the presumed Neo-Babylonian *GU Text*.

Other drawings in Mesopotamian art have been identified as drawings of constellations as well. Among these are representations of a bird and a fish with their tails entangled. This has been identified as the Babylonian constellations "The Swallow" and Anunitum, which come together to form the classical/modern constellation Pisces, thus matching what is written in D i 4–9:[15]

D Section A, col. i 4–9 (see p. 39)

i 4. [The star which] stands opposite "The Field," (is) "The Swallow":

i 5. (It is) a bird, a star with wings, flying, that is it has wings.

i 6. The constellation which stands after "The Field" is Anunitu, a river.

i 7. ["Th]e Swallow" and Anunitu at their tails cross one another,

i 8. the stars of the Tigris and Euphrates, by the crest of "The Tail"

i 9. they are held together. "The Swallow" and the stretched out neck of Anunitu, the station of Venus.

Such descriptions of the constellations in our group often seem to go far beyond what might be constructed from matching stars in the sky with elements of constellations. For example, the very complex description of the constellation ^{mul}UD.KA.DUH.A, "The Demon with The Open Mouth," with its two faces, one human and one that of a lion, in D i 17-26; the complex picture of the astronomical goddess Gula sitting on a chair by her dog ("The Dog" = Hercules) in ABC Sections

---

[14] See below, p. 12.

[15] For such drawings, see Kurtik (2007, 738–9), with discussion on pp. 44–45.

IX–X; and the image of "The Old Man," his staff, and his sheep in E Section A. Thus, our group sometimes describes constellations as they were actually seen in the sky by the human eye, and other times how the constellations were rendered into images of gods, animals, and objects in the sky in the minds and artwork of ancient Mesopotamians.

With this in mind, we can perhaps speak of two different types of descriptions of constellations in ancient Mesopotamia: First, there is the one in which constellations and/or their parts are defined by the number of their stars and these stars' configuration, thus yielding a set of points and connecting lines, such as those presented on the planisphere CT 33 10 and the Sippar Planisphere edited in Horowitz and Al-Rawi (2001). Such descriptions in our group include the numerous references to the heads of constellations as a single star;[16] the description of the Big Dipper as a four-sided geometric figure *apsamakku* with a wagon pole consisting of three stars;[17] and the description of the eleven stars of "The Crab" in D Section F, the latter of which provides an excellent match to the drawing of this same constellation on the Sippar Planisphere. However, in many other places, our group provides descriptions of constellations and constellation parts that go far beyond sets of points and lines. For example, the aforementioned description of the two faces of ^{mul}UD.KA.DUH.A, "The Demon with the Open Mouth," in D Section B, provides a much more graphic view of the constellations head and face than what might be gained from identifying a constellation's head with one star only. Likewise, the description of the tails of "The Swallow" and Anunitu crossing one another in D Section A, and the many statements that constellations are in dressed human form, flesh out, so to speak, images of Mesopotamian constellations in the sky that are not unlike the intricate flowing figures of the classical constellations with which we are familiar from our own civilization or, for that matter, those we find in Late Babylonia in the drawings of constellations on the micro-zodiac texts.

One might suspect that such images and drawings may in fact lie behind the occasional references to buildings being decorated with stars and other heavenly designs in the cuneiform corpus, and the Mesopotamian concept of the *šiṭir šamê*, "heavenly writing," but this cannot be proved on the basis of currently available evidence.[18]

THE URANOLOGY TEXTS

Our group of tablets containing uranology texts numbers five. Of these, the only one to be previously published is our Source A, VAT 9428, a tablet in Neo-Assyrian script from Assur, edited by E. Weidner in 1927.[19] When complete, Source A offered descriptions of the drawings of constellation in twelve sections of text.

[16] Outside of our group, references to more than one star in the head of a constellation are more common. For example, "The Two Stars of the Head of 'The Lion'" appear in *ziqpu*-star lists, and the heads of "The Hired Man" and "The Scorpion" are divided into two and three stars, respectively, when used as Normal Stars. Cf. also the three stars in the forehead (sag-ki = *pūtu*), but not the head (sag = *rēšu*), of Aries in D i 15.

[17] ABC Section VII, D iii 13–16 (Section L).

[18] For this term and discussion, see Rochberg-Halton (2004a, 1–2); Horowitz (2011, 228).

[19] A very brief discussion of A is also available in Hunger and Pingree (1999, 65) and in White (2007, 276).

Relatively complete sections survive for "The Old Man," "The Great Twins" and "The Lesser Twins," "The Crab" (Cancer), "The Lion" (Leo), a constellation in human form representing the goddess Eru, "The Wagon" and "The Wagon of Heaven" (the Big and Little Dippers), and another constellation representing the goddess Gula. Two fragmentary near duplicates of this text can now be identified. The first is a British Museum tablet in Babylonian script, BM 66958, probably from Babylon, which gives part of the text known from VAT 9428 on one of its sides. This will be our Source B, with the relevant side labeled Side A. Second is the Late Babylonian Yale fragment NBC 7831 from Uruk, our Source C, which now preserves only one side of its original. Source B duplicates part of the reverse of A, and C part of A's obverse, but B and C themselves do not overlap. On the basis of orthography, B seems earlier than C: B is written in a hand typical of the Neo-Babylonian or early Persian period, whereas C might be better placed in the late Persian or early Hellenistic period.

For the most part, the texts of ABC are very close to each other, although here and there one finds some minor variances,[20] the most important of which is a disagreement between A and C regarding "The Lesser Twins": C giving each of "The Lesser Twins" a *kurkurru*,[21] whereas A does not. This already suggests that the earlier Source A dating to the Neo-Assyrian period, and B and C in Babylonian script, are less-than-exact duplicates, a supposition that can also be confirmed on the basis of differences in line division between A and BC. For example, in Sections II–IV, A uses one more line of text than C, and the same is true for A and B in Sections VII and IX. These differences, however, are more stylistic than significant in terms of contents. Thus, for all intents and purposes, the three manuscripts function as three separate witnesses for the same text, and so we may refer to VAT 9428//BM 69958 Side A//NBC 7831 as ABC.[22] The only exception to this rule is found at the very top of the current fragment C where this manuscript adds two lines before the beginning of A. These two lines may not be part of the main text of our ABC, but rather from an introduction written later, or something that proceeded ABC in the mind of the scribe of C.

Our Source D is another Late Babylonian Yale tablet, MLC 1866.[23] Its colophon indicates that is from the Bit Reš Temple in Uruk and dates to the 97th year of the Seleucid Era (215–4 B.C.E.). D is by far the largest and longest tablet in our group, and the most complex, even though what survives of D gives only the upper half of the original manuscript. Much of Source D gives the same type of material as ABC. In fact, Source D, Sections C, D, F, G, I, J, and N, take more or less the same form as the descriptions of constellations in Sources ABC. Yet, other parts of D offer much more than our Sources ABC, both in quantity and scope of interest.

[20] For example, in Section II the forward "Twin" of "The Great Twins" carries a *hinšu* in his left hand, but in C (as well as D, also from late-period Uruk) it is placed in his right hand. Note also a slightly different sequence of wording in Section III (A 9//C 9') in regard to the two stars of the heads of "The Lesser Twins." For *hinšu*, see below, p. 49.

[21] For *kurkurru*, see below, pp. 78–79.

[22] Side B of B does not belong to our group but is astronomical. For an edition and discussion, see Appendix B. As noted above, the reverse of C is lacking.

[23] Beaulieu and Britton previously provided information on D that was used for a brief notice in Hunger and Pingree (1999, 63), where this tablet is discussed along with A on ibid., 65. Kurtik (2007) quotes some passages from the text as well.

Source D is written over six columns. Our text occupies col. i–iv, with the surviving part of col. v giving its colophon. Col. vi preserves the remains of an illegible text belonging to something else, and so does not apparently belong to our group. The obverse and the opening lines of D col. iv include descriptions of the drawings of constellations as in ABC, but also related information that seems to flow from the stylus of its composer in almost free association. For example, in the first section of the text which begins with a description of "The Field," we find, among other digressions, discussion of the Pisces as a fish and bird formed from the Mesopotamian constellations Anunitu and *Šinūnūtu*, "The Swallow" (D i 4–9), and then an exegesis on the name of the Babylonian constellation "The Hired Man," which justifies how the same stars that comprise this constellation in human form in Babylonian tradition can be Aries in the form of a sheep (D i 12–15).

A number of other passages in D show an interest in connecting the constellations to the patron deities of the Bit Reš, the heaven god Anu and his wife Antu, reflecting their supreme importance in the late Uruk pantheon.[24] Yet, despite such differences between ABC and D, one can still recognize that both belong to a group of texts intended to describe constellations.

This is not the case for the last part of D col. iv, consisting of Sections O and P (iv 7'–14' and 15'–20'). The first line of Section O does give what looks like the beginning of a description of "The Fish" (Piscis Austrinus), but by the time we reach the next line we find that we are in a discussion of the planets and their relationship to Anu, Antu, and their vizier Papsukkal. Section P continues in the same vein as Section O, here associating various Mesopotamian twins constellations with Anu and Papsukkal. Thus, Sections O and P seem to be of a different order than either ABC or what comes earlier in D. This can perhaps be justified on astronomical grounds for the planets in Section O, because the planets emit only one point of light and so, unlike constellations, cannot be described in terms of parts and paraphernalia. This, however, is not the case for the "The Twins" constellations discussed in Section P. In fact, D Sections C, D, E offer descriptions of "Twins" constellations of the type found in ABC Section II–III. Thus, we suggest that Sections O and P have a different agenda than ABC and most of D. We suggest that these two sections are not meant to describe constellations but to stress the connections between what is visible in the sky, and Anu and his household in the Bit Reš; in other words, to connect Anu's realm in Heaven with his realm on Earth. True, this is also a consideration in the earlier part of D, but much more space there is allocated to describing how to draw constellations than to explaining their relationship to Anu.

The differences between the simple group ABC and the more intricate materials in Source D can be illustrated by a comparison of sample selections: below, the description of a drawing of "The Wagon" (the Big Dipper) from the ABC subgroup, and then two passages from D. The first of these is the aforementioned discussion of the Babylonian constellation "The Hired Man" from the first part of D, where the main interest still seems to be to describe constellations. The second

[24] The upper edge of D, D ii 10, 21, D iv 20, and D v 4'–5' in the colophon. In contrast, Anu and Antu are not mentioned in ABC. For further discussion, see below, pp. 58–59.

is from Section O in the latter portion of Source D, where the primary concern is Anu and his household:

ABC Section VII (see p. 25)

¶ "The Wagon" is an *apsamakku.*[25] [4 s]tars are drawn at its fore.
Its pole (is) towards the heel of Eru. 3 stars
on its pole – 1 bright star at the head of the pole
and 2 lower stars side by side <in front> on the pole – are drawn.

D Section A, col. i 12–16 (see p. 39)

i 12. . . . . the star which stands after it is
i 13. "The Hired Sheep Man" (Aries = ^{mul}LÚ.HUN.GÁ), "The Hired Sheep." Nisan is the month of Anu.
i 14. It is the star of the New Year. "The Hired Man" (Aries), (it is) a lamb, a ram,
i 15. the . . of "The Hired Sheep" – three stars are drawn at its forehead;
i 16. two stars are drawn on its thigh: four stars stand at its feet.

D Section O, col. iv 8'–14' (see p. 42)

iv 8'. ¶ Jupiter, Venus, Me[rcury, Satur]n;
iv 9'. ¶ The planet Mars, the Moon and the Sun.
iv 10'. Seven gods, sons of Anu, who by the seed of Anu
iv 11'. were begotten, the Igigi-gods. Pa[psukk]al (is)
iv 12'. the counselor of Anu. The seven of them in the abode of Anu,
iv 13'. the king, magnificently they s[tan]d, a rival
iv 14'. they have not.

The final tablet in our group, MLC 1884, is our Source E. This is yet another Late Babylonian fragment from Yale; its slanting script suggests it is from very late in the history of cuneiform writing, perhaps as much as a century or more younger than the Uruk tablets that we know as Sources C and D. Source E stands on its own, apart from both D and the group ABC. It does not duplicate any portion of ABC or D, and, unlike the other tablets, is concerned with only two groups of constellations rather than a series of constellations that form a full cycle of the sky. However, like D, E does make use of entries from *Mul.Apin* and uses similar terminology and phraseology in its descriptions of the constellations. As in the case of B, on E our text occupies only one side of the tablet, here the reverse, with what is preserved divided into two sections: Section 1 begins with a description of a constellation whose name is not fully preserved but given what comes next is most likely "The Old Man" (^{mul}ŠU.GI), which is described here in terms familiar from ABC;

[25] For *apsamakku*, see below, p. 31.

for example, dressed, bearded, and set with a *kurkurru*. The section then continues with what appear to be parts of this constellation including a crook that "The Old Man" holds; this is none other than the constellation ^{mul}GÀM, "The Crook" itself. Section 2 is less well preserved than Section 1. The first few lines can be restored with near certainly. Here we find references to three constellations—"The Pleiades," "The Bull of Heaven," and "The True Shepherd of Heaven"—which occur in the same sequence in *MulApin* I i 44–ii 2. The following lines are more difficult to follow, but we suggest below that they refer to the moon in its *bīt niṣirti* in Taurus. The obverse of E is not obviously astronomical or astrology and so does not belong to our group.

THE SOURCES

A. VAT 9428

Photographs: plates 2–3[26]
Copy: Weidner (1927, 74–5)
Previous Edition: Weidner (1927)
Neo-Assyrian, Assur
Single Column, both obverse and reverse belong to our group

The 1927 copy by Weidner is generally accurate. Only minor discrepancies can be observed, such as slight deviations in the relative position of signs, and a tendency by Weidner to draw standard forms of Neo-Assyrian signs rather than to reproduce what is actually present on the tablet. There may also be a few instances of what may be overzealous restoration of signs in breaks or damaged portions of the tablet, but such discrepancies between what we see today and what Weidner copied are more likely to be the result of physical deterioration of the tablet. Weidner's copy is, after all, now more than 80 years old. More misleading is that Weidner's copy does not give an accurate sense of the physical format of the tablet—that the tablet is inscribed on four sides (obverse, reverse, upper and lower edge).

The tablet is divided into twelve sections that are divided from one another by single horizontal rulings:

Obverse	Sections 1–4
Lower Edge	Section 5
Reverse	Sections 6–10
Upper Edge	Sections 11–12

B. BM 66958 (= 82-9-18, 6952) Side A

Photographs: plate 4
Copy: plate 5
Neo-/Late Babylonian, Babylon
Single Column, only Side A belongs to our group

[26] The photograph numbers for VAT 9428 in Berlin are VAN 4075–4077, 4954–4958.

This fragment, which probably originates from Babylon,[27] contains our text on one side (side A) and a collection of planetary observations on the other side (side B, see Appendix B). The style and terminology of the reports of planetary observations are very similar to that found in three fragments believed to be from Nineveh (two written in Assyrian and one in Babylonian script) and likely dating to the seventh century B.C.E.[28] This raises the possibility that both sides of BM 66958 were copied from Assyrian sources.

C. NBC 7831

Photographs: plate 6
Copy: plate 7
Late Babylonian, Uruk
Fragment, the preserved side belongs to our group, the second side is lacking

D. MLC 1866

Photographs: plates 8–9
Copy: plates 10–14
Late Babylonian, Uruk
Seleucid Era 97 (4 January 214 B.C.E.)

Three columns each side, both obverse and reverse col. i–ii belong to our group. Reverse col iii seems to belong to something else. The colophon in rev. ii marks the end of the portion of the tablet belonging to our group.

E. MLC 1884

Photographs: plate 15
Copy: plates 16–17
Late Babylonian, Uruk
Single Column

Our text is given on the reverse of E. The obverse is not astronomical; see Appendix C.

STRUCTURE OF THE TEXTS AND THE REPERTOIRE OF CONSTELLATIONS

The texts in both the simple group ABC and the expanded group D and E have essentially the same structure. Each section begins with a DIŠ sign used as an item marker followed by the name of a constellation which is then described. References to other nearby constellations may then be given. These may either be simple

[27]The 82-9-18 collection contains tablets shipped in twelve cases, eleven of which came from Sippar and one from Babylon, plus a few tablets found at Dailem (ancient Dilbat) (Reade 1986, xxxiii). The collection contains about 35 astronomical fragments including one, BM 65831, that joins BM 34575+, a System A lunar ephemeris from Babylon (Steele 2006b, Text D), and several Astronomical Diaries and Normal Star Almanacs of the Seleucid Period, which must also be from Babylon. In the absence of other evidence, it is therefore most likely that all the astronomical texts in this collection were found at Babylon.
[28]Pingree and Reiner (1975).

references to constellations that appear in the neighboring sections of the text (often used for reference in the description of the first constellation; e.g., by saying that the constellation faces toward the second constellation) or longer descriptions of constellations not found elsewhere in the text. Thus, we have two types of constellations in our texts: constellations that govern a section and are marked as items using the DIŠ sign, which we will call primary constellations; and secondary constellations within a section, which are unmarked. In some sections, a planet or the moon (and probably the sun, although no example is preserved) is also given as a secondary entry in the section. These references to the moon and planets do not relate to the planets themselves, which of course move through the constellations, but rather to the positions of their *bīt niṣirti*, "secret place," a position in the sky of particular astrological significance for that body. The *bīt niṣirti* is a forerunner of the Greek astrological doctrine of planetary exaltations (hypsoma) and is referred to already in *Enūma Anu Enlil* and in various Neo-Assyrian royal inscriptions as well as in later astrological cuneiform texts including the Horoscopes.[29] The signs of the zodiac (and hence approximately the constellations) in which the *bīt niṣirti* are located are as follows:

Sun:	Aries
Moon:	Taurus
Mercury:	Virgo
Venus:	Pisces
Mars:	Capricorn
Jupiter:	Cancer
Saturn:	Libra

The descriptions of constellations in the group follow certain general patterns. Following the name of the constellation, a general description of the shape and form of the constellation is given—whether it is a human figure, an animal, or an inanimate object (such as part of a wagon); its general shape; and, for human figures, whether it is clothed, bearded (for male figures), and if it has a *kukkuru*. There follows a detailed description of the parts of the constellation. In the case of human figures, there is a strong tendency to describe elements from above to below, whereas for animals and objects the tendency is to describe elements from front to back. In the case of right and left, what is on the right is always described before what is on the left.

There is good reason to believe that the repertoire of constellations in our group ultimately derives from *MulApin*. The first tablet of *MulApin* contains several (overlapping) star lists including a list of the stars in the Paths of Enlil, Anu, and Ea; a list of *ziqpu*-stars; and a list of the constellations in the Path of the Moon (i.e., the zodiacal constellations). All the constellations in our group are found in *MulApin* with the exception of ^{mul} ^{d}*Gula*, "Stellar Gula," of ABC Section IX–X. However, this constellation probably included parts of the constellation ^{mul}ÙZ, "The She-Goat," who is associated with the goddess Gula in *MulApin* I i 24. D and E quote directly from the lists of the constellations in the three paths.

[29] Rochberg-Halton (1988, 53–7).

TABLE 1. Primary and secondary entries in each section of ABC

Section	Primary Entry	*Mul.Apin* Catalogue	Secondary Entry	*Mul.Apin* Catalogue
I	Old Man	Enlil 3	The Stars	Anu 5
II	Great Twins	Enlil 5	The Jaw of the Bull	Anu 6
III	Lesser Twins	Enlil 6	(none)	(none)
IV	The Crab	Enlil 7	Jupiter The Lion	Enlil 33 (planet) Enlil 8
V	The Lion	Enlil 8	(none)	(none)
VI	Eru	Enlil 10	The Lion	Enlil 8
VII	The Wagon	Enlil 14	Eru	Enlil 10
VIII	The Wagon of Heaven	Enlil 18	The Wagon	Enlil 14
IX	The Dog	Enlil 23	Stellar Gula Sitting Stars Standing Stars	? Enlil 22 Enlil 20
X	Stellar Gula	?	(none)	(none)
XI	[. . .]	[. . .]	[. . .]	[. . .]
XII	[. . .]	[. . .]	[. . .]	[. . .]

The simple group, ABC, is divided into twelve sections. The primary and secondary constellations given in ABC are listed in table 1 along with the place of the constellation within the lists of constellations in the three paths from *Mul.Apin*. It is clear that the primary constellations are an ordered subset of the Enlil constellations from *Mul.Apin*. Unfortunately, Sections XI and XII are too badly damaged to identify the constellations described in those sections, but it seems likely that these constellations would have completed a full circuit of the Enlil-stars. We suggest, therefore, that the twelve sections may reflect the notion of month-stars of the type found in the Astrolabe tradition which lists one star from each of the three stellar paths for each month of the year; here in ABC, twelve Enlil-stars.[30] This raises the question of whether ABC represents one of three original compositions containing twelve constellations for each of the three different paths.

The secondary entries in ABC appear mostly as reference points for describing the primary constellation. Some of the secondary constellations are taken from the neighboring Path of Anu and so are next to but below (south of) the primary constellations. An allusion to the Path of Anu is also found in A rev. 13. As noted previously, the reference to Jupiter relates to its *bīt niṣirti*, which is known to be in Cancer. The statement that Jupiter is between "The Crab" and "The Lion" is in agreement with the illustration on the obverse of the micro-zodiac text VAT 7847 + AO 6448.

The structure of the entries in D is somewhat more complicated than in ABC. As discussed previously, only col. i–iii and the first part of col. iv (Sections A–N) contain the constellation description text. Only about half of the height of D is preserved,

[30] Note, however, that the stellar repertoire in ABC is not drawn from Astrolabes or any other related text that gives stars in groups of twelve. In particular, note that the names of "The Twins" constellations given in ABC are ^{mul}MAŠ.TAB.BA.GAL.GAL and ^{mul}MAŠ.TAB.BA.TUR.TUR as in *Mul.Apin*, rather than ^{mul}MAŠ.TAB.BA and ^{mul}MAŠ.TAB.BA.GAL.GAL as in the Astrolabe tradition, and that the list of constellations that head each section does not agree with any of the Astrolabe texts.

which implies that there are missing sections at the end of col. i, ii, and iii and at the beginning of col. iv. Given the great variability in the length of the preserved sections (e.g., A extends over sixteen lines and B over at least nine lines, whereas G and H are each only two lines in length), it is not possible to accurately estimate how many sections are missing in each break: A rough estimate based purely on the size of the tablet would suggest somewhere between two and five missing sections between B and C and between H and I, and anywhere between four and ten missing sections between L and M, giving a total of between eight and twenty missing sections.

The primary and secondary entries in each section are presented in table 2.

Unlike ABC, the sequence of primary entries in D is not restricted to one of the three paths. Instead, these entries in D move backward and forward between the paths, approximately, but not precisely, in accordance with the movement of the sun during the course of the year. Thus, they more or less follow the paths of the moon, sun, and planets through the zodiac. The secondary entries are again nearby stars to

TABLE 2. Primary and secondary entries in each section of D

Section	Primary Entry	*Mul.Apin* Catalogue	Secondary Entry	*Mul.Apin* Catalogue
A	The Field	Anu 1	The Swallow Anunitu Venus The Hired Man/ Sheep	Anu 2 Anu 3 Anu 20 (planet) Anu 4
B	The Demon with the Open Mouth (2–5 sections missing)	Enlil 26	The Pig [The Horse]	Enlil 27 Enlil 28
C	[The Great Twins]	Enlil 5	(none)	(none)
D	The Lesser Twins	Enlil 6	The True Shepherd of Anu	Anu 7
E	The True Shepherd of Anu	Anu 7	The Twins, Lulal and Latarak The Rooster	Anu 8 Anu 9
F	The Crab	Enlil 7	[Jupiter]	Enlil 33 (planet)
G	The Arrow	Anu 10	The Crab	Enlil 7
H	The Bow (2–5 sections missing)	Anu 11	(none)	(none)
I	The Snake	Anu 12	The Raven	Anu 13
J	Ninmah	Ea 4	(none)	(none)
K	EN.TE.NA.BAR. HUM	Ea 5	Šullat and Haniš [Numušda]	Ea 7 Ea 8
L	The Wagon (4–10 sections missing)	Enlil 14	The Crook The Fox The Ewe [The Hitched Yoke]	Enlil 4 Enlil 15 Enlil 16 Enlil 17
M	[. . .]		(none)	(none)
N	[The Great One]	Ea 2	(none)	(none)

TABLE 3. Primary and secondary entries in each section of E

Section	Primary Entry	*Mul.Apin* Catalogue	Secondary Entry	*Mul.Apin* Catalogue
1	The Old Man	Enlil 3	The Crook The Wagon	Enlil 4 Enlil 15
2	The Stars	Anu 5	The Bull of Heaven/ The Jaw of the Bull Moon	Anu 6 Planet

the primary entry. Mostly, these come from the same path as the primary entry, but sometimes they come from the neighboring path—the same as we found for ABC. Once more, the planets appear in the position of their *bīt niṣirti*.

If we are correct in restoring "The Great One" as the primary entry in Section N, the final section, then the text apparently gives constellations through a full circuit of the sky. Thus, the text describes the constellations around the sky roughly following the order of the path of the sun (i.e., following the zodiac).

D also differs from ABC in quoting constellation names with their epitaphs directly from the lists of Enlil, Anu, and Ea constellations found in *Mul.Apin* (with some modifications to reflect the pantheon of Seleucid Uruk). This characteristic is also found in E. E contains only two or possibly three sections if there is a missing section below the break. The entries in E are summarized in table 3.

Interestingly, "The Old Man" and "The Stars" appear together as primary and secondary entries in Section I of ABC, but in E they each have their own sections with their own secondary entries. Because "The Old Man" is an Enlil-star and "The Stars" is an Anu star in *Mul.Apin*, it is tempting to suggest that E originally contained three sections (the third section lost in the break below Section 2), each concerning a primary entry for one of the three paths. If our interpretation of line 17' is correct as referring to the moon, this would agree with the position of the moon's *bīt niṣirti*, which is placed between "The Stars" and "The Bull of Heaven" on the micro-zodiac text VAT 7851.

A synopsis of the descriptions of the constellations by alphabetical order of the ancient name as it appears in the group, and a list of parts and elements that comprise the constellations, is found in Appendix A. There we also list parallels to the descriptions in our group from sources such as the *ziqpu*-star texts, *The GU Text*, *The DAL.BA.AN.NA Text*, and the Normal Star lists, and, where appropriate, we provide suggested modern equivalences to the ancient Mesopotamian stars and their parts.

STATEMENTS OF DIRECTION

Many of the descriptions in our texts include references to the placement of the stars or features of a constellations using either absolute (e.g., right or left) or relative (e.g., in front of, behind, below) statements of direction. Most of these terms are familiar from other astronomical texts and appear to hold the same meaning here. For example, in astronomical contexts the directions "in front of" and "behind" refer to the order with which a celestial object crosses a fixed point in its daily cycle of rotation, such as rising over the eastern horizon, transiting the meridian, and setting at the western horizon. Objects that are "in front of" another object cross these points first, whereas objects that are "behind" cross afterward. This usage is extended to apply to the position of moving objects—that is, the sun, moon, and planets—relative to stars close to the ecliptic. The moon, for example, can be stated to be so many cubits or fingers in front of a particular star on a given occasion; the next day, the moon may then be so many cubits behind the same star as the moon moves forward through the zodiac, passing the star and going beyond it. Before reaching the star, the moon will rise before the star; after passing the star, the moon will rise after or behind the star. Thus, "in front of" and "behind" refer both to the order with which objects rise or set and also to the relative position of two objects in the zodiac: an object moving with direct motion through the zodiac is in front of another object when it has not yet reached it, and behind once it has. Astronomically, this means that for objects near the ecliptic, "in front of" refers to lower celestial longitude and "behind" refers to higher celestial longitude.

With one possible exception, all of the references to "in front of" and "behind" in our texts agree with the standard use of these terms in astronomical texts. Furthermore, the order of entries in all of our texts progresses, just like in *Mul.Apin*, from stars that rise first (i.e., those that are in front) to those that rise later (i.e., those that are behind). Within entries such as those for the various twins constellations, the front twin is always given before the rear twin, following the general principle that objects in front rise or set first. Thus, the terms "in front of" and "behind" refer to the daily rotation of the heavens rather than to, for example, the direction in which a constellation is facing. The only possible exception to this rule is in references to the position of Jupiter between the Crab and the Lion. For discussions of these entries in ABC and D, see the commentaries to those tablets.

The directions "right" and "left" in our texts are in all cases compatible with the idea that these are the right and left sides of the constellation from its own perspective as it faces us. Thus, for constellations fairly near the ecliptic, "right" corresponds to lower celestial longitude and "left" to greater longitude. This is confirmed by the order of the entries within sections, which go either from right to left or from front to back, and also by some individual entries such as ABC Section VI in which Eru is holding a whip in its right hand which stretches out over the Lion. Unless the whip goes over Eru's head, it must stretch to lower longitude and so the right side of the body must be at lower longitude.

These understandings of statements of direction in our texts have significance for how we understand both the descriptions and the drawings of the constellations on the micro-zodiac texts. As has long been known, the drawings on the micro-zodiac texts are the mirror-image of what we see if we look at the night sky. The reason, simply put, is that these illustrations, like the descriptions in our texts, follow the order of rising across the eastern horizon. What rises first appears in first place on the illustration—that is, on the left of the drawing—just as it appears in first place in the descriptions. By placing the front of the illustration to the left, the sun moves through the illustration from left to right; thus, as we move through the text we move through the year, which is exactly what is being done in the text written below the illustration. The illustrations, therefore, are not to be visualized as looking in on the celestial sphere from the outside, as some have suggested, but rather are a consequence of the illustration following the direction of writing on a tablet from left to right.

TOWARD A HISTORY OF THE GROUP

It is not currently possible to write a complete history of the group. The earliest source, Source A, is from Neo-Assyrian period Assur. The material on ABC is clearly related to several other astronomical texts that are attested for the first time in Neo-Assyrian sources, including *Mul.Apin*, the Astrolabe planispheres, the *ziqpu*-star texts, and what we know as *The DAL.BA.AN.NA Text*. *Mul.Apin*, like our Source A, is in fact first attested in manuscripts from Assur and as we demonstrated above, the repertoire and sequence of stars in ABC follows that for the Enlil-stars in *Mul.Apin*. Furthermore, the *ziqpu*-star texts and *The DAL.BA.AN.NA Text* make numerous references to parts of constellations that match those in our group. This is also true of *The GU Text*, which is roughly contemporary to the later part of the Neo-Assyrian period, but so far only attested on a tablet from Babylonia.[31] Thus, we find that our group shares features with a wide set of astronomical texts that were in circulation during the Neo-Assyrian period; this raises the possibility that our group ultimately grows out of the astronomical traditions of the Neo-Assyrian empire. If ABC has its origins in Assyria, then our Babylonian Sources B and C, which nearly duplicate A, could indicate that the text then came to be known and copied in the south, both in central Babylonia in the middle of the first millennium as is the case for Source B, and then later in Uruk for Source C.[32] However, based on the current evidence, or rather lack of evidence, it is also possible to argue that the group was Babylonian in origin and was transmitted to Assyria where Source A was prepared, and thus our Sources B and C were survivors of an original Babylonian tradition.

The history of D and E presents even more of a problem. As discussed previously, D shows close affinities to *Mul.Apin* in col. i–iii and the top of col. iv, both in its quoting of entries from *Mul.Apin* and in the sequence of groups of stars considered in the surviving text. Similar affinities to *Mul.Apin* are found in E, but what is left on E does not include the types of discussions, digressions, and exegesis that characterize D. In this, Source E is closer to ABC than D, so it is clear that D and E do not together form a subgroup on the order of ABC. Thus, we do not have the convenient situation of a set of simpler manuscripts ABC which form the basis for a more complex D/DE. Furthermore, ABC and D, and also D and E, do not share a common stellar repertoire. Yet, the continued interest in D and E in describing constellations suggests that the two sources are ultimately derived from a tradition represented by ABC. Thus, we suggest that the authors of both D and E built their texts on the basis of predecessors of the type known from ABC. If we assume ABC once had sister tablets for constellations in the Paths of Anu and Ea, we can imagine a Neo-Assyrian period set of three tablets, each with descriptions of twelve primary constellations of the type of Source A, thus yielding a grand total of 36 constellations and descriptions—12 each for each of the three stellar paths. This, of course, is the same number of stars as in the Astrolabe tradition, including the Neo-Assyrian period Astrolabe planispheres (CT 33 10–11), the Astrolabe text CT 33 9,[33] and also *Mul.Apin* I ii 36–iii 12, which lists 36 stars that rise in sequence

[31] In this context it is worth noting the possible Assyrian origin of the material on the other side of B; see Appendix B.
[32] On the survival of Neo-Assyrian texts in late-Uruk tablet collections, see Beaulieu (2010).
[33] Horowitz Alb (chapter 13, 209–24).

over a 360-day ideal astronomical year.[34] The author(s) of what eventually became Source D and E could, in theory at least, have made use of the descriptions of constellations on all three of our posited tablets to complete their work, but of course they could have added additional stars as well. Unfortunately, no evidence for sister tablets to ABC is now at hand to support the hypothesis outlined previously.

The interest in Anu and Antu in D suggests that D, as it has reached us, was composed by an adherent of Anu and Antu, almost certainly in Uruk after the fall of Babylon to the Persians in 539 B.C.E., because this when the rise of Anu and Antu becomes pronounced in tablets from Uruk. But what was the source for this material? One possibility is that the author of what we now know as D developed new material on his own, perhaps on the basis of existing learned oral traditions of the type sometimes found in texts identified as *šut pî* (oral traditions) or *mukallimtu* (learned traditions). A parallel to such a process within the realm of astronomy and astrology may be found in the accretion of commentary and other materials to the core omen collections of the *Enūma Anu Enlil* tradition over the course of the first millennium. If so, it may be instructive that the colophon of D tells us that the scribe of our tablet comes from the family line of a registrar of *Enūma Anu Enlil*.

Another set of influences may be sought in the background to religious and astronomical speculation and exegesis of the type found in many of the texts published in Livingstone (1986). Similar materials relating specifically to astronomy and astrology are available in "the Great Star List" and the Astrolabe *mukallimtu* BM 82923, which often give unexpected identifications between stars and deities (i.e., not according to the tradition of Astrolabes and *Mul.Apin*).[35] Thus, Source D may have one foot in two different parts of the cuneiform astronomical tradition. Like BM 82923, which is both an Astrolabe text and a *mukallimtu*, D may have its roots in *Mul.Apin*-type astronomy and simple descriptions of the type we find in ABC, but in its final form, it may be closer to the realm of what Livingstone (1986) calls mystical-explanatory texts. In any case, the colophon dating D to Seleucid era Year 97 serves as a terminus pro quem for the completion of the Uranology portion of D in its current form.

The type of tablets on which our text is preserved also deserves comment. It is clear that the text in ABC was a standard text that was known over a wide geographical and chronological span.[36] Source B shows that the ABC text was sometimes copied by a scribe onto a tablet along with other astronomical material, and Source C may have done so as well. Similarly, both Sources D and E were copied onto tablets alongside other texts. These multi-text tablets suggest that our texts were not copied simply for the sake of copying or as part of the scribal curriculum, but rather were copied by active astronomical scribes. This practice of scribes copying different texts onto one tablet is well known in the corpus of Babylonian astronomical texts. For example, one particular scribe compiled several different texts all dealing with either *ziqpu*-stars or Normal Stars on the tablet BM 36609+, and he also compiled a range of different astrological texts concerning the calendar, stars, and the zodiac onto another tablet, BM 36303+.[37] The scribe of LBAT 1501 copied a *ziqpu*-star text, a description of the upper and lower limits of the

[34] Mul.Apin I ii 36–iii 12, see Hunger and Pingree (1999, 65–6).

[35] Horowitz Alb (chapter 7.3, 139–56), with an edition of the related text Sm. 1492, which makes mention of Adapa as does the colophon to Source D.

[36] On the concept of "standard texts" in the corpus of Babylonian astronomical tablets, see Steele (2014).

[37] For BM 36609+ see Roughton et al. (2004). For BM 36303+ see Steele (2015).

path of the moon, and a scheme for the latitude of the moon onto his tablet; the same description of the moon's path and the lunar latitude text are preserved on LBAT 1502 along with a different *ziqpu*-star text.[38]

A comparison of the physical format of D with another compendium tablet, the Seleucid period Astrolabe anthology LBAT 1499 from Babylon, can help to better understand the layout of our tablet. LBAT 1499 contains two distinct separate units of textual material, what we will call LBAT 1499 Part I and Part II, which are separated by approximately 10 lines of vacant space on the reverse of the tablet. LBAT 1499 Part I (Sections 1–4) is directly related to the Astrolabe tradition.[39] Part II is not. Instead, Part II (Sections 5 and 6) gives material related to the rising times of the zodiac of the type that makes use of some of the Astrolabe stars.[40] This can explain why Part I and Part II came to be recorded on the same tablet. Thus, LBAT 1499 can be thought of as a "scribal workbook" composed of two parts: the Astrolabe anthology as Part I, and the rising time materials as Part II. If this model can be applied to D, we can think of the uranology text and its colophon in D col. i–v as MLC 1866 Part I, and the short work of up to 35–40 lines in col. vi as MLC 1866 Part II. In this case, the colophon at the end of Part I is not the colophon of MLC 1866 as a whole, but is instead only the colophon of Part I; that is, the colophon of a previous tablet that was copied onto MLC 1866, which would explain the anomaly of MLC 1866 giving a column of text after a colophon. It would also suggest that our Source D was relatively popular, because we would now have indirect evidence for three copies:

1) MLC 1866 Part I
2) the tablet that MLC 1866 Part I was copied from, with its colophon indicating that Part I was copied from a previous source[41]
3) this previous source

In LBAT 1499, and the other tablets discussed previously, there is clearly a common theme to the different texts that have been copied together onto a single tablet. If MLC 1866 follows this model, it would suggest that the text given in MLC 1866 vi was in some way related to the uranology group, at least in the eyes of the scribe of MLC 1866.

We suggest that our text MLC 1866 and the other thematically unified compilation tablets discussed previously[42] were written as what we have called "scribal workbooks"—tablets on which learned scribes copied materials from previous sources for their own use, not unlike the notebooks (or computer files) that we, the

[38] On the lunar latitude scheme, which is a section from the latitude scheme found on Atypical Text E, see Steele (2012). On the texts describing the path of the moon, see Steele (2007).

[39] A list of the 36 Astrolabe stars with values for the length of day and night that parallels Alb B Section I, followed by three sections of omens which are unattested elsewhere in the Astrolabe tradition: one section of omens for the Ea-stars, a section of omens for the Anu-stars, and a section of omens for the Enlil-stars.

[40] The rising time scheme is discussed by Rochberg-Halton (2004b) and Steele (2017a).

[41] D v 3'. This would also mean that the date given in the colophon is not necessarily the date of MLC 1866, which could have been inscribed substantially later than its source material. However, the colophon that dates D to Seleucid Year 97 would still serve as a *terminus pro quem* for the completion of the Uranology portion of D (MLC 1866 Part I) in its current form. Cf. the commentary to D col. i for the use of the *Glossenkeil* there as another indication that D Part I was copied from previous source material.

[42] Another astronomical member of this genre might be the Astrolabe anthology BM 55502, for which see Horowitz Alb (26–7).

modern Assyriologists, use to transcribe cuneiform tablets in our own research.[43] If this is correct, it explains why MLC 1866 as a whole (col. i–vi) and LBAT 1499 do not have colophons. Colophons are meant for future readers. We suggest that scribes copied onto their own workbook tablets material that, for the time being, was intended for their own eyes only, with (or without) future intent to recopy such material—most likely with colophons—for future use by others.

Finally, there is the matter of the correspondence between some of the descriptions of constellations in our group and the drawings on the micro-zodiac tablets. This gives rise to a suspicion that our group is describing drawings of constellations that were known to its scribes, rather than/or in addition to observed configurations of the constellations in the sky. If we are correct in this conclusion, it might suggest that the drawings on the micro-zodiac tablets may go back to earlier drawings of the same type. Although only three such micro-zodiac tablets are known to us, there are additional micro-zodiac tablets from Babylon which include the labels that would appear on the drawings but omit the drawings themselves.[44] The presence of these labels suggests that a complete set of twelve drawings existed and was known at both Babylon and Uruk. Could these drawings have been influenced by *Mul.Apin*, which dates to this early period and makes reference to parts of constellations? Was there once a full set of drawings of the constellations of *Mul.Apin*? If so, is this why the verb used for drawing the constellations in our group is *eṣēru*, "to draw," as in drawing a picture? Furthermore, might such sets of drawings be the ultimate source of at least some of the descriptions of constellations in D and E, rather than just a text tradition of the type known to us from ABC?

It is worth noting here that our texts are essentially descriptive rather than procedural. Procedural or instructive texts are always written in the second person ("you draw") and often begin with a statement of purpose ("In order for you to . . ."), whereas the texts presented here are all written in the third person ("is drawn") and do not begin with a statement of purpose.[45] Thus, they are describing the appearance of either constellations or drawings of constellations rather than providing instructions on how the reader should draw them. Of course, it is possible that a reader could use these texts as a guide to make such drawings, but it suggests that this was not the primary aim of the text. Rather, these texts fall into the category of descriptive astronomical texts, akin to the various Greek texts that describe constellations, such as Aratus's *Phaenomena*, which, for our group at least, may be seen as a precursor.

To conclude, we must admit that at present we know very little about the genesis and history of our group, although we can say that the group must have been popular, at least in limited circles. This popularity allowed for its transmission from the Neo-Assyrian to the Seleucid period and for the appearance of sources from at least three cities: Assur, Babylon, and Uruk.

---

[43] One may also compare the scrolls that medieval scribes used when copying from earlier manuscripts for further study, and perhaps how Mesopotamian scribes might have sometimes used wax tablets and writing boards.

[44] See, e.g., BM 42288+42644+43414+43716 (Monroe 2016).

[45] Compare this with the texts LBAT 1494 and 1495, which describe the construction of some type of gnomon instrument. These texts begin with a statement of purpose—"for you to make"—and provide second-person instructions for drawing lines on a tablet or brick, which acts as the base of the gnomon. Instead of the third person verb *e-ṣir*, "is drawn," which is used in our texts, LBAT 1494 and 1495 use the second person *te-eṣ-ṣir*, "you draw."

CHAPTER 2

The Simple Group

Edition

A. VAT 9428. Photograph: plates 2–3; copy: Weidner (1927, 74–5). Previous edition: Weidner (1927).[1]
B. BM 66958, Side A. Photograph: plate 4; copy: plate 5.[2]
C. NBC 7831. Photograph: plate 6; copy: plate 7.

Section I

A 1. ⸢¶ ^{mul}ŠU.GI⸣ *ṣal-mu lu-bu-uš-tu*$_4$ *z[i-iq-na za-qi]n*
A 2. x x x x *i-na* ZAG-*šú* *[dir-rat na]-ši*
A 3. GÙB-šú ina UGU MUL.MUL LÁ-*át ina* SAG-[*šú* 1 MUL] ⸢*e*⸣-*ṣir*

C 3'.[3] . . ZA]G-⸢*šú di*⸣-*ra*-[*at*]
C 4'. . . .] 1 MUL *e-ṣ[ir]*

Section II

A 4. ¶ ^{mul}MAŠ.TAB.BA.GAL.GAL 2 *ṣal-mu zi-i[q-na zaq-nu kur-ku-r]a šak-nu*
A 5. 1-*n[u]*-⸢*ú*⸣ MUL.MEŠ *ina* S[AG-*šú-nu e*]*ṣ-ru*
A 6. *ṣal-mu* IGI-*ú šá* IGI ^{mul}*Is le-e h[i]-in-ši ina* ⸢ŠU⸣ G[ÙB]-⸢*šú*⸣ *na-ši*
A 7. *ṣal-mu* EGIR-*ú* U$_4$.SAKAR *pa-a-ša ina* ŠU GÙB-*šú* [*n*]*a-ši*

C 5'. . . . ⸢2 *ṣal*⸣-[*mu* x x x x *ku*]*r-kur-ra šak-nu* 1-*nu-ú* M[UL . . .
C 6'. . . . AL]AM IGI-⸢*ú šá* IGI⸣ ^{mul}*Is le-e hi-in-šú ina* ŠU$_{II}$ 15-*šú na-ši*
C 7'. . . . EGIR]-⸢*ú*⸣ U$_4$.SAKAR *pa-a-šu ina* ŠU GÙB-*šú na-ši*

---

[1] The transliteration of A below is based on Weidner's copy, the photographs, and collations by W. Horowitz. Weidner's sign-forms are accurate, but his positioning of signs is sometimes not reliable. The tablet today is for the most part as Weidner saw it, although here and there some of what Weidner copied is no longer visible.

[2] Side B does not belong to our group. For an edition of Side B, see Appendix B.

[3] C 1'–2' form part of a section that ends with a horizontal ruling. We can read:

C 1': . . .] x hu x
C 2': . . . -*t*]*um e*-[*ṣir*]

The end of C 2', with the form of *eṣēru* restored, could be part of an introduction to C, or it could form part of a sort of ABC Section 0, giving a description of a constellation that is not in A or B. The sign HU in line 1 could be for MUŠEN, but if so it is too far to the right to be part of the name of the constellation that is being described even if C 1' is the first line of the section. In any case, such speculations must remain only that until more text of our group is recovered.

Section III

A 8. ¶ ^{mul}MAŠ.TAB.BA.T[UR.TUR] ⸢2⸣ *ṣal-mu lu-[b]u-uš-tu*$_4$ *ziq-na zaq-*⸢*nu*⸣
A 9. ⸢2 MUL.MEŠ *ina*⸣ [SAG-*šú-nu*] *eṣ-ru ṣal-mu* IGI-*ú ina* ŠU ZÀG-*šú*
il-tuh-a na-ši
A 10. x x x *d*[*ir-rat it-t*]*i il-tuh-a ṣa-bit*
A 11. *ṣal-m*[*u*] ⸢EGIR⸣-[*ú* ZAG-*šú*] *ana* IGI-*šú bir-qa na-šat*
A 12. ŠU G[ÙB-*šú bir-qu ana šá-ša*]*l-li* [*ṣa*]*l-*⸢*mi*⸣ *ši-i ú-kal*

C 8'. [¶ ^{mul}MAŠ.TAB.BA.TUR.TUR] 2 *ṣal-mu lu-bu-uš-tu*$_4$ *ziq-nu* ⸢*za-aq*⸣*-nu*
ku-ur-ku-ra šá-a[*k-nu* . . .
C 9'. [*ina* SAG-*šú-nu* 2 MU]L.MEŠ *eṣ-ru* ALAM IGI-*ú ina* ŠU$_{\text{II}}$ 15-*šú*
il-tah-ha na-⸢*ši*⸣
C 10'. [. . . *di-ra-at it*]*-ti il-tah-ha ṣa-bit* ALAM EGIR-*ú ina* ŠU$_{\text{II}}$ 15-*šú bír-q*[*a* x x]
C 11'. [. . . *bi*]*r-qu ana šá-šal-li* ALAM *ši-i* [*ú-kal*]

Section IV

A 13. ¶ ^{mul}AL.LUL *áp-sà-*[*m*]*a-ak-*[*ku* 4] MUL.MEŠ *ina i-tu-ti-šá eṣ-ru*
A 14. *i-na* ŠÀ*-bi-šú-nu* [6 MUL.MEŠ *a-hi-i*]*n-nu-ú*[4] *a-ha-meš rak-bu*
A 15. 1 MUL *ina* SAG-*šú* ⸢*e*⸣-[*ṣir* ^{mul}S]AG.ME.GAR *ina* IGI-*šú e-ṣir*
A 16. ^{mul}UR.GU.LA *a-n*[*a* EGIR-*át* ^{mul}SA]G.ME.GAR *e-ṣir*

C 12'. [¶ ^{mul}AL.LUL *á*]*p-sà-ma-ak* 4 MUL.MEŠ *ina i-tu-ti-šú eṣ-r*[*u*]
C 13'. [x x x *a*]*-ha-meš rak-bu* 1 MUL *ina* SAG-*šú e-ṣir* [. . .
C 14'. . . .] *ana* EGIR ^{mu}[^{l}SAG.ME.GAR . . .

Section V (Lower Edge)[5]

A 17. ¶ ^{mul}UR.MAH SIPA *šar-h*[*u* x x x x x] UGU KUN-*šú* MUL *ana* IGI *hu-ru-u*[*p-p*]*i*
A 18. *ù qu-ru-*⸢*ub*$^?$⸣ 4$^?$ M[UL$^?$.MEŠ$^?$ x x 1 MU]L *ina* GABA-*šú e-ṣir*

C 15'. [. . . SA]G$^?$.GAR.ME$^?$ GUB-*zu*

REVERSE

Section VI

A r1. ¶ ^{mul}E$_4$.RU$_6$ *ṣal-m*[*u lu-bu-uš-t*]*u*$_4$ *kur-ku-ra* GAR-*in*
A r2. 1 MUL *ina* SAG-*šú e-*[*ṣir qi*]*n-n*[*a-za ina* Š]U ZAG-*šú na-ši dir-rat*
qin-na-zi-šú
A r3. *ina* UGU KUN ^{mul}UR.GU.LA L[Á*-á*]*t ina* ŠU GÙB-*šá* MUL *na-ši*

B 1'. ⸢¶ ^{mul}⸣ [. . .
B 2'. [*q*]*in-na-zi-š*[*u* . . .

---

[4] Restoration on the basis of D ii 19. For the word, see CAD A$_{\text{I}}$ 184, *ahennā, ahinnū*, "each separately."
[5] For a proposal for the place of what remains in C in respect to A, see the commentary.

Section VII

A r4. ¶ ^{mul}MAR.GÍD.DA *áp-sà-ma-ak-ku* [4 M]UL.MEŠ *ina pu-ut-ti-šá eṣ-ru*
A r5. *ma-šad-da-šá ana šá-šal-li šá* ^{mul}⸢*E*$_4$⸣*-ru*$_6$ 3 MUL.MEŠ
A r6. *ina* IGI*-at ma-šad-di-šá* 1 MUL *na-bu-u ina* SAG *ma-šad-di*
A r7. *ù* 2 MUL.MEŠ *šap-lu-tu*$_4$ Á *ana* Á*-šú ina* <<SAG>>[6] *ma-šad-di eṣ-ru*

B 3'. ¶ ^{mul}MAR.GÍD.DA *áp-sà-*⸢*ma*⸣*-*[*ak-ku* . . .
B 4'. *šá* ^{mul}*E*$_4$*-ru*$_6$ 3 MU[L.MEŠ . . .
B 5'. *šap-lu-tu*$_4$ Á *ana* Á*-šú i*[*na* . . .

Section VIII

A r8. ¶ ^{mul}MAR.GÍD.DA.AN.NA *ka-lak-ku a-na* IGI *ma-šad-di šá* ^{mul}MAR.GID.D[A]
A r9. *ù ma-šad-da-šá* IGI *ka-lak-ku šá* ^{mul}MAR.⸢GÍD.DA *i*⸣*-ma-šá-a*[*l*]

B 6'. ¶ ^{mul}MAR.GÍD.DA.AN.NA *k*[*a-lak-ku* . . .
B 7'. *u ma-šad-di-šá ana* IGI *k*[*a-lak-ku* [. . .

Section IX

A r10. ¶ ^{mul}UR.GI$_7$ *kal-bu šá ina* UGU *ur-ki-ti-šú eṣ-r*[*u* IGI*-šu ana* $^{mul\ d}$*G*]*u-*⸢*la šak-nu*⸣
A r11. 2 MUL.MEŠ GABA-su *ina* KUN*-šú* 7 MUL.MEŠ [*eṣ-ru ina l*]*i-mi-tu*$_4$
A r12. 9 MUL.MEŠ AN.TUŠ.A.MEŠ 3 MUL.MEŠ *n*[*a-bu-ti* MUL.MEŠ[7] A]N.GUB.BA.MEŠ[8]
A r13. *ù* 6 MUL.MEŠ *bit sak-ki-i i-ti* K[ASKAL? MUL.MEŠ *š*]*u-ut* ^{d}*A-nim li-me-tam*

B 8'. ¶ ^{mul}UR.GI$_7$ *kal-bu šá* UGU [. . .
B 9'. 2 MUL.MEŠ GABA*-su ina* K[UN*-šú* . . .
B 10'. [A]N.⸢TUŠ.A.MEŠ 3⸣[9] MU[L.MEŠ . . .[10]

Section X

A r14. ⸢¶ ^{mul}⸣ ^{d}*Gu-la ṣal-mu lu-bu-uš-tu*$_4$ *in*[*a* SAG]*-šá* 1 MUL *e-ṣir*
A r15. [*ina* Š]U ZAG*-šá* MUL *na-šat* ŠU GÙB*-šá* x [x x] x GAR*-át*
A r16. [2 MUL.MEŠ] Á *ana* Á *šap-la-an k*[*ab-li* ^{giš}G]U.ZA *eṣ-ru*

---

[6] For the emendation, see CAD M$_1$ 351. Without the emendation, the line is nonsense, with one star at the head of the pole and then two more stars at the head of the pole. One may suspect an error of dittography.

[7] There may not be enough room to restore both MUL and MEŠ, although this would provide a symmetry between the names for "The Sitting Stars" and "The Standing Stars."

[8] For the readings AN.TUŠ.A.MEŠ and AN.GUB.BA.MEŠ and their Akkadian renderings *antušû* and *angubbû*, see the commentary.

[9] Just the top of the last vertical of the numeral survives.

[10] End of B.

Section XI (Upper Edge)

17. [¶ ^{mul} . . . . . .] *li* x[11] [. . . .]-⸢*nim-ma*⸣
18. [. . . . .] NI [. . .

Section XII (Upper Edge)

19. [¶ ^{mul} . . .
20. [. . . . .] GABA [. . .
21. [. . . . 2 MUL].MEŠ Á *ana* Á [. . .

--- [12]

TRANSLATION

Section I

1. ¶ "The Old Man" is a clothed human figure w[ith a bear]d,
2. . . . on his right hand a las[h he car]ries.
3. His left (hand) extends above "The Stars." At [his] head [a star] is drawn.

Section II

4. ¶ "The Great Twins" are two human figures, with b[eards], set with a [*ku*]*rkurru*.
5. One star apiece are [d]rawn at [their] h[ead].
6. The front figure, which is before "The Jaw of the Bull," carries a *hinšu*[13] in his l[eft][14] hand.
7. The back figure carries a (crescent-shaped) sickle-axe in his left hand.

Section III

8. ¶ "The L[esser] Twins" are two clothed human figures with beards, s[et] with a *kurkurru*.[15]
9. 2 stars are drawn [at their heads]. The front figure carries a whip in his right hand.
10. . . . a l[ash wi]th the whip he holds.
11. The back figure, his right hand carries a lightning-bolt.
12. [His] l[eft] hand holds a lightning-bolt to the back of this figure.

Section IV

13. ¶ "The Crab" is a four-sided figure. 4 stars are drawn at its circumference.
14. Inside them [6 stars, side by] side straddle one another,

---

[11] The sign is not ME and so the word here cannot be *limītum* as in Section IX.
[12] Ruling at the very end of the upper edge followed immediately by the obverse.
[13] For *hinšu*, see the commentary to D ii 1.
[14] C: right hand.
[15] C only.

15. 1 star at its head is drawn. [J]upiter is drawn in front of it.
16. "The Lion" to the rear of [Ju]piter is drawn.

Section V (Lower Edge)

17. ¶ "The Lion" is a magnifice[nt] shepherd [. . . . .] above its tail a star in front of the *di[s]h/pel[v]is*[16]
18. and *nearby*? *4*? s[*tars*? . . (with) 1 sta]r drawn in its chest.

REVERSE

Section VI

1. ¶ Eru is [a clothe]d human figu[re,] set with a *kurkurru*.
2. 1 star is d[rawn] at its head, it carries [a w]hi[p in] its right [ha]nd. The lash of its whip
3. over the tail of "The Lion" str[etch]es out. In her left hand it carries a star.[17]

Section VII

4. ¶ "The Wagon" is an *apsamakku*. [4 s]tars are drawn at its fore.
5. Its pole (is) towards the heel of Eru. 3 stars
6. on its pole – 1 bright star at the head of the pole
7. and 2 lower stars side by side <in front> on the pole – are drawn.

Section VIII

8. ¶ "The Wagon of Heaven" is a wagon box before the pole of "The Wago[n],"
9. and its pole before the wagon box of "The Wagon" in a similar manne[r.]

Section IX

10. ¶ "The Dog" is a dog which is draw[n] (sitting) on its haunches. [Its face] is set [toward Stellar G]ula.
11. 2 stars (are) its chest, in its tail 7 stars [are drawn. Along the cir]cumference
12. 9 are "The Sitting Stars" – 3 are the b[right] stars of "The S]tanding [Stars"],
13. and 6 are stars of *Bit Sakki* adjacent (to) *the P*[*ath* of the stars o]f Anu, (but) outside.

---

[16] So A as preserved. For C and a very tentative proposed restoration for the first part of the line, see the commentary to C 15'.

[17] Note the masculine form of the stative for *ṣalmu* as opposed to a feminine form for the goddess.

Section X

14. ¶ Stellar Gula is a clothed human figure. A[t] her [head] 1 star is drawn.
15. [in] her right [ha]nd she carries a star. Her left hand . [. .] . is set
16. [2 stars] side by side below the l[egs of the c]hair are drawn.

Section XI (Upper Edge)

17. [¶ The (star) . . . . . .] . . [. . . .] . .
18. [. . . . .] . [. . .

---[18]

Section XII (Upper Edge)

19. [¶ The (star) . . .
20. [. . . . . . .] chest [. . .
21. [. . . . 2 star]s side by side [. . .

---[19]

Commentary[20]

Section I: A 1–3, C 4', ^{mul}ŠU.GI = *šību*, "The Old Man"

The human form of ^{mul}šu-gi = *šību*, "The Old Man" (Perseus), is explicit in the text and implicit in the constellation's ancient name. In our text, the constellation is clothed and bearded and carries an item now missing in the break in his right hand. His left hand extends above "The Stars." One star is drawn at his head. A number of body parts of the constellation are also known elsewhere: Its feet in the astronomical report and letter SAA 8 216 and SAA 10 100;[21] chest and feet in the omens BPO 2 72, XV 9–10 and 74, XVI 5, 7; and its *kinṣu* and *asīdu*, "shin" and "knee," in TCL 6 18+, 15.[22] In SAA 8 380, the constellation has a *kurkurru* as do a number of other stars in human form in our group.[23]

The head of the constellation is listed last among its body parts, after (according to our restoration) its right and left hands. This appears to go against the convention of listing body parts of constellations in human form from above to below, but it may be explained in that the something belonging to the constellation extends above "The Stars" in line 3 and so may have been raised over and above the constellation's head. If so, we suggest that this element corresponds to the sword of Perseus which the classical constellation holds in his left hand above his head.

A 1: We have here the first use in our group of Akkadian *ṣalmu* with the technical sense of constellation in human form, a meaning which is demonstrated

---

[18] The ruling is still visible in places, although not indicated on Weidner's copy in *Archiv für Orientforschung* 4 (see the photograph, plate 5).

[19] End A. Ruling at the bottom of the upper edge followed immediately by the obverse.

[20] The notes follow the order of A.

[21] See also the commentary in Parpola 1983, 308–9 (LAS 300 + 110 = SAA 10 100).

[22] Edition Weidner (1925). See also Hunger and Pingree (1999, 274); Kurtik (2007, 484, 490).

[23] For the other examples of the term *kurkurru* and discussion, see p. 31.

time and again.[24] In A, this term is written syllabically.[25] In C, the term is also written syllabically (C 5', 8') at the start of the entries for the two "Twins" constellations, but with the sumerogram ALAM when referring to just one of the figures (C 6', 9'–11'). Akkadian *ṣalmu* is also used numerous times with this same meaning in D (see the commentary to D i 18–19), and in E 2'.

A 2: The missing element that the constellation holds in its hand is a lash (*dirratu*), which can be restored from C 3'.

Section II: A 4–7, C 5'–7', ^{mul}MAŠ.TAB.BA.GAL.GAL = *tū'amu rabûtu*, "The Great Twins"

"The Great Twins" is one of three sets of "Twins" constellations which appear in *Mul.Apin*, the others being "The Lesser Twins," which like "The Great Twins" is in the Path of Enlil, and "The Twins Which Stand Opposite the True Shepherd of Anu: Lulal and Latarak" in the Path of Anu. "The Great Twins" are clearly the most important of these "Twins" constellations as they appear not only in the list of constellations in the three paths but also, unlike "The Lesser Twins" and "Lulal and Latarak," in the list of *ziqpu*-stars and in the list of constellations in the Path of the Moon (i.e., the zodiacal constellations). It is almost certain, therefore, that it is "The Great Twins" that should be identified with the zodiacal constellation the "Twins" which is commonly used in the Late Babylonian period, eventually giving its name to the zodiacal sign, and from which stars were used as Normal Stars.

There is a small discrepancy between A and C. In the Neo-Assyrian Source A, the forward "Twin" carries a *hinšu* in the left hand,[26] but in C it is placed in his right hand. This is also the case in the parallel portion of D ii 1–3 (D Section C). "The Twins" constellations are also considered in D iv 16'–20'. In all sources, "The Twins" constellations include a forward and rear twin.

A 5: The writing 1-*nu-ú* for *ištēnû*, with the numeral and phonetic complements, is similar to a number of those given in CAD I/J 279 for the word. The intent here is to stress that each of the two Gemini of "The Twins" has its own separate head.

A 7: CAD P 268 *pāšu* d takes the ax and crescent here to be two separate items. We follow Weidner (1927, 76) who takes what the rear twin is holding as only one item, a *Sichel-Axt*, our (crescent-shaped) sickle-axe.[27]

Section III: A 8–12, C 8'–11', ^{mul}MAŠ.TAB.BA.TUR.TUR, "The Lesser Twins"

This section parallels that for "The Lesser Twins" in D ii 4–9. As with "The Great Twins," "The Lesser Twins" too include a front and back "Twin." Again, there are some small discrepancies between Neo-Assyrian A and Late Babylonian C and D from Uruk; for example, the inclusion of a *kurkurru* in C 8' and D 6', but not in A.

---

[24] Cf. a near parallel with *ṣalmu* and the D-stem of *eṣeru* in Gilg. I i 49, there with reference to the body of Gilgamesh: *ṣalam pagrīšu bēlet-ili uṣir*, Belet-Ili designed the (human) form of his body (George 2003, 540).

[25] A: 1, 4, 6–9, 11, rev. 1, 14.

[26] The meaning of *hinšu* is uncertain. See the discussion in the commentary to D ii 1 (Section C), p. 49.

[27] See Wallenfels (1993, 283) for the rear figure of a Gemini holding a sickle-axe.

A 9: Note the different sequence of wording in A 9 and C 9'.

A 10: For the restoration here and in C 10', see the corresponding phrase in D ii 8: *dir-rat it-ti il-tuh-hu ṣa-bit*. The word *iltuhhu* for "whip" is replaced by *qinnazu* in Section VI. The two are synonyms according to *Urgud* and the commentary to *Ludlul-bel-nemeqi* (see CAD I/J 288 *ištuhhu*).

A 11: The restoration of Weidner (1927, 75) is confirmed in part by C, but with the restored right hand of the constellation in A supplying the subject of the grammatically feminine stative *našât*. Again, here note slightly different wording between A and C.

A 12: The reading in C 11' now allows us restore A 12. We understand this to mean that the back "Twin" holds a lightning-bolt in front of himself with his right hand, and lightning-bolt behind himself with his left hand, making *ana šašalli*, in effect, a double preposition of the type "*ina muhhi*," "*ina libbi*," "*ana pāni*," and so forth. If the lightning-bolt was actually to be located on, or at the *šašallu*, "tendon of the heel," we might have expected the prepositions *ina* or *itti* (on, with). Note the use of the present-future form *ukāl* rather than a form in the stative as in the previous lines.

C 8': This line of C provides "The Lesser Twins" with *kurkurru*, but this is not the case in A. Thus, in A, only "The Great Twins" have *kurkurru*, whereas in C both sets of twins have this feature.

Section IV: A 13–16, C 12'–14', ^{mul}AL.LUL, "The Crab"

Like "The Great Twins," "The Crab" appears in all three major star lists in *Mul. Apin*: the list of stars in the three paths, the list of *ziqpu*-stars, and the list of constellations in the Path of the Moon. In the Late Babylonian period, "The Crab" appears regularly as a zodiacal constellation, eventually giving its name to a sign of the zodiac. Four Normal Stars are associated with "The Crab."

The first two lines of this section parallel D ii 18–19 (Section F) which also consider "The Crab." The commentary to the passage in D on pp. 50–51 offers discussion of the arrangement of the eleven stars that comprise the constellation.

A 13: The four stars that form the points of the *apsamakku* in "The Crab" are used as Normal Stars in the Astronomical Diaries and related texts where they are called "The Front Star of the Crab to the North," "The Front Star of the Crab to the South," "The Rear Star of the Crab to the North," and "The Rear Star of the Crab to the South," and are to be identified as η, θ, δ, and γ Cancri, respectively.

A 14: For *rakābu* with the sense "to straddle," see CAD R 87 4. For the verb in astronomical contexts, see ibid., 88 4. Cf. the writing U_5-*u'* for *rabkū* in the parallel entry in D ii 19.

A 15–16: ABC Section IV places Jupiter in relation to "The Crab" and "The Lion" in two consecutive statements:

[J]upiter is drawn in front of it ("The Crab")

"The Lion" to the rear of [Ju]piter is drawn.

Jupiter's *bīt niṣirti* is located between "The Crab" and "The Lion." The stars that make up the head of "The Lion" are at the lowest celestial longitude of the stars in that constellation. "The Lion," therefore, is facing toward "The Crab," with Jupiter in between the two; the drawing on the obverse of the micro-zodiac tablet VAT 7847 + AO 6448 confirms that "The Lion" faces toward Jupiter. In normal astronomical usage, therefore, we would expect the following positional relationships: "The Crab" is in front of Jupiter = Jupiter is behind "The Crab" and Jupiter is in front of "The Lion" = "The Lion" is behind/to the rear of Jupiter. Our text, however, although it gives the expected positional relationship between "The Lion" and Jupiter, has the opposite of what we would expect for "The Crab" and Jupiter. We can think of two possible explanations for this. First, it is possible that if "The Crab" is facing toward Jupiter and "The Lion" (which is implied by the ordering of stars in the description and is also the case for the Greek Cancer), the author of our text may have intended "in front of" to be read literally in the sense of looking forward from the point of view of "The Crab." This would run counter to all other known cases of the use of this terminology, however. Second, it is possible that scribe has made a simple mistake here.

For "The Lion" and Jupiter, see also the commentary to Section V immediately below.

Section V: A 17–18, C 15', ^{mul}UR.MAH, "The Lion"

"The Lion" also appears in the three primary star lists of *Mul.Apin* and in the Late Babylonian period it gave its name to a sign of the zodiac. In *The GU Text*, "The Lion" has two stars at its head, four stars in its chest, a star for its right front foot, a star for a foot in the middle of the constellation, two stars for its rump, and a single star in its tail;[28] with the exception of the star in the Lion's foot, all of these stars/star groups also appear in the standard list of 25 *ziqpu*-stars.[29] Five Normal Stars are also known from the Lion: "The Head of the Lion" (ε Leonis), "The King" (α Leonis), "The Small Star Which is 4 Cubits Behind the King" (ρ Leonis), "The Rump of the Lion" (θ Leonis), and "The Rear Foot of the Lion" (β Virginis).

The description of the constellation is broken in A, and almost completely lacking in C (no parallel is preserved in D). What does survive speaks of the constellation's tail, its *huruppu* (partially restored), four stars in an uncertain context, as well as a star, or stars, belonging to the constellation's chest. A drawing of the constellation is available on the obverse of the micro-zodiac text VAT 7847 + AO 6448. In the drawing we see more detail including its head, mane, four legs, a mark on his thigh, and his tail. A sort of shadow of the "The Lion" follows the main constellation. The reason for this shadow is uncertain. It may be simply a trace left over from a first attempt by the artist to draw "The Lion," which the artist placed by mistake too far to the right.[30]

A 17: Note the rendering of the name of "The Lion," ^{mul}UR.MAH, rather than ^{mul}UR.GU.LA as in A 16 above and A rev. 3 below: Cf. SAA 8 45: 5–6 for a likely restoration ^{mul}UR.MAH = ^{mul}[UR.GU.LA], and SAA 8 437 (^{mul}UR.MAH and ^{mul}UR.GU.LA in consecutive omens). Also note Urra XXII 253' for variant entries: ^{mul}UR.MA[H = *Ne*$^{?}$-*e*$^{?}$-*šú*$^{?}$, ^{mul}UR.GU.⸢LA⸣ = *La*-⸢*ab*$^{?}$-*bu*$^{?}$⸣. The identification of ^{mul}UR.MAH, "The Lion," as a shepherd can be compared with *Sm.* 1492 rev.

[28] Hunger and Pingree (1999, 98).

[29] Steele (2014).

[30] There may be a similar preliminary drawing of Virgo, about an inch to the right of the actual drawing—but this may just be a trick of the light in the published photographs.

3 where ^{mul}UR.GU.LA, "The Lion," is identified with the goddess Baba, who herself is identified there as a shepherdess:[31]

> ^{mul}UR.GU.LA ^{d}*Ba-ba*$_{6}^{!}$ (Text TA) MUNUS.SIP[A . . .
>
> "The Lion," Baba, the shepherde[ss . . .

The *huruppu* is a body part of quadrupeds including cattle and now lions. We can identify this feature of "The Lion" with that drawn near the animal's thigh in the drawing on VAT 7847 + AO 6448. This word appears in the dictionaries as *huruppu*: a metal dish, and by extension a part of the body and cut of meat (see *AHw* 360, CAD H 256). CDA 122 offers a translation for the body part: "pelvic basin" of an ox.[32] It seems likely that the *huruppu* is the round (dish-shaped) element on the thigh of "The Lion," above its back legs, in the image preserved on the VAT 7847 + AO 6448. The *huruppu* also occurs as part of a constellation in LBAT 1513 iii 4': . . . *ziq-pi hu-ru-pi-šú*.[33] The *huruppu* of "The Lion" probably corresponds to the Normal Star "The Rump of the Lion" (θ Leonis).

A 18: This feature with four stars is almost certainly to be identified with the four stars in the chest of the "The Lion" in *The GU Text* and *ziqpu*-star texts.[34]

C 15': The reading here is most uncertain. One possibility is that the scribe mistakenly wrote [^{mul}SA]G$^{?}$.GAR.ME$^{?}$ for ^{mul}SAG.ME.GAR "Jupiter." This planet is placed in proximity to "The Lion" (written ^{mul}UR.GU.LA) in the previous line at the end of ABC Section IV ("The Lion" to the rear of Jupiter). If this reading is correct, we may propose that A 17//C 15', originally read as follows:

> ^{mul}UR.MAH SIPA *šar-h*[*u*] *a-n*[*a* EGIR-*át* ^{mul}SA]G.ME.GAR GUB-*zu*
> UGU KUN-*šú* . . .
>
> "The Lion" (is) a magnifice[nt] shepherd that stands t[o the rear of Ju]piter.
> Above its tail . . .

However, it must be stressed that the reading of both [SA]G$^{?}$ at the start of the present line and ME, rather than MAŠ/BAR, are not secure.[35]

Section VI: A rev. 1–3, B 1'–2', ^{mul}E$_{4}$.RU$_{6}$, Eru

For the reading of the star-name note the gloss *e-ru* in one of the exemplars of *Mul.Apin* I i 11.[36] The feminine possessive pronoun *ša* with GABA and 3FS stative in rev. 3 reflect the gender of the goddess Eru.[37] Elsewhere in the entry, the 3MS *naši* and suffix -*šu* refer back to *ṣalmu*. LBAT 1510 8'–11' reports that Eru holds a

---

[31] See Horowitz Alb (52–3).
[32] This same part of the animal, ^{uzu}*hu-ru-up-pu*, is also attested on a tablet from Al-Yahudu (Johannes and Lemaire 1999, 2, 10) with a new edition by C. Wunsch expected soon.
[33] The writing in LBAT 1513 with PI rather than BI, as well as a number of writings cited in the dictionaries with PA rather than BA, make it clear that the last consonant is normally P.
[34] Hunger and Pingree (1999, 85–91). For nine stars in "The Lion" constellation in the *ziqpu*-star text AO 6478, see ibid., 85.
[35] A reading with *šá-maš* may also be contemplated, but we have no reason to expect the Sun in the context of "The Lion."
[36] Hunger and Pingree (1989, 21).
[37] GAR = *šá* cannot be read here as the numeral 4 here if for no other reason than that the next sign is MUL, for a single star, rather than MUL.MEŠ for multiple stars.

sissinnu, "date frond," in the right hand (see CAD S 326 *sissinnu* 4). $^{mul}E_4.RU_6$ also appears in the *ziqpu*-star lists.

Section VII–VIII: A rev. 4–7, 8–9; B 3'–5', 6'–7', ^{mul}MAR.GÍD.DA = *eriqqu*, "The Wagon"; ^{mul}MAR.GÍD.DA.AN.NA = *eriqqi šamê*, "The Wagon of Heaven"

Both "The Wagon" (part of Ursa Major) in Section VII and "The Wagon of Heaven" (part of Ursa Minor) in Section VIII are composed of two elements: a pole (*mašaddu*), and the main part of the wagon which is formed by four stars set in a rectangular-type pattern. In Section VII the four stars are said to form the *apsamakku* of "The Wagon." This is also the case in D iii 14. In Section VIII this feature of both wagons is described as a *kalakku*. Thus, our text does not fully distinguish between the two shapes. Robson (1999, 50–4) studies the two shapes in mathematical texts, there concluding that the *apsamakku* is a concave square, whereas the KI.LÁ = *kalakku* is a square truncated pyramid (Robson 1999, 101; earlier Neugebauer and Sachs 1945, 65); both being four-sided figures. "The Crab" is also said to have an *apsamakku* in AC Section 4 and D ii 18.

The remainder of the discussion of "The Wagon" in D iii 13–23 differs substantially from that in Sources AB. After D iii 23, the column breaks off. It is most likely that the next portion of D, when complete, was for "The Wagon of Heaven," but this cannot now be demonstrated, particularly since no subsequent section of Source D matches anything that comes later in ABC.

A rev. 4–7: The Akkadian name of the constellation, *eriqqu*, is a feminine noun with a plural *eriqqātu*, hence the feminine possessive pronouns in A. The feminine grammatical gender matches the constellation's identification with goddesses including Ninlil as in *Mul.Apin* I i 15 (¶ ^{mul}MAR.GÍD.DA ^{d}*Nin-líl*), and the star-catalogue of Alb B (partially restored).[38]

A rev. 8: For the name of the constellation in Akkadian, *eriqqi šamê/šamāmi*, "The Wagon of Heaven," see Horowitz (1989).

A rev. 9: The traces at the end of the line seem to admit nothing other than this present-future form of *mašālu*. Cf. the end of Section III, A 12 with the present-future *ukāl*. Note also the feminine pronoun here as with "The Wagon" above. Cf. ^{mul}MAR.GÍD.DA.AN.NA identified with the goddess Damkianna in both *Mul. Apin* I i 20 and LBAT 1513 rev. i 9'.

Section IX: A rev. 10–13, B 8'–10', $^{mul}UR.GI_7$ = *kalbu*, "The Dog"

No parallel section survives on D.

This constellation consists of the most stars of any in our group; as many as 27 consisting of the constellation's chest and tail, and divided into three clusters: "The Sitting Stars," "The Standing Stars," and the stars of *Bit Sakki*.

First are the nine main stars that apparently give the basic shape of the constellation: two in the chest and seven in its tail. Then, there are two more groups of nine: the nine "Sitting Stars" (*antušû*), which in turn are followed by another set of

[38] Alb B I iii 10: ^{mul}MAR.GÍD.DA ^{d}*Nin-*[*líl*]. So too our restoration of D iii 13. See also the Astrolabe text BM 82923 15: ^{mul}M[AR.GÍD.D]A 50 *be-el-tum*, "The W[ago]n," 50, The Lady (Horowitz Alb 141–2). For further identifications with goddesses, see ibid., 116.

nine—the three "Standing Stars" (*angubbû*) plus the six stars of *Bit Sakki*. The two groups, "The Sitting Stars" and "The Standing Stars," appear elsewhere,[39] but this is the only attestation of the *Bit Sakki* group of stars.

A rev. 10: We restore the double-determinative for the stellar goddess Gula on the basis of Section X immediately below, with the masculine plural form of the stative at the end of the line supporting the restoration IGI = *pānû*, "face" (a masculine plural noun). Thus, what we seem to have here is the image of the dog-constellation (Heracles) sitting on its haunches facing the stellar goddess Gula in human/divine form, just as Gula's dog is often depicted looking upward at the goddess in Mesopotamian art.

A rev. 12: We choose the readings AN.TUŠ.A.MEŠ (= *antušû*) and AN.GUB. BA.MEŠ (= *angubbû*) for the two star-groups on the basis of a late writing ^{d}AN. GUB.BA.MEŠ in the mystical work AO 17626: 12.[40] Here, a reading *^{d}*Dingirgubbû*, with consecutive DINGIR-signs (AN.AN) would be very awkward. Once the reading GUB, "to stand," is established, TUŠ, the opposite, "to sit" seems appropriate. BA assures the reading GUB for the DU-sign here.

A rev. 13: CAD S 78–79 takes the name of the star-group from *sakkû* B, but this is the one and only attestation of this word. Thus, given the long writing *sak-ki-i* one is tempted to take the word from *sakkû* A, "rites, ritual regulations"; that is, translating the name of the group as "The House of the Rites," or even from the word *sakkû* C, a type of headgear, that is attested but once, in the lexical list An.

The second half of the line is unclear due to the break. Our tentative restoration *h[arrān kakkabāni š]ūt* ^{d}*Anim* is based on the shape of the very badly damaged piece of sign just before the break which gives an impression of being the beginning of KASKAL (= *harrānu*), the length of the break, and what survives toward the end of the line. This restoration also allows us to understand *limētam* at the end of the line as a writing for an adverbial form of *limītu*, here with ME as is attested in Neo-Assyrian, but without a preceding preposition or following noun as is normally the case when the word is used in this way. In other words, if our interpretation is correct, the *Bit Sakki* stars are in the Path of Anu, but along the border between this path and the Path of Enlil. "The Dog" is an Enlil-star in *Mul. Apin* I i 25.

Section X: A rev. 14–16, $^{mul\ d}$Gula, Stellar Gula

No parallel section survives on D.

In Section IX (restored) and Section X we have the first certain attestation of the star-name Stellar Gula ($^{mul\ d}$*Gu-la*), a constellation in the form of the goddess, as opposed to the constellation ^{mul}GU.LA, "The Great One" (Aquarius), which is precluded by the feminine pronouns and stative forms in A rev. 15 for the goddess. As noted in the discussion of Section IX above, Stellar Gula appears with her stellar dog, ^{mul}UR.GI$_7$, thus placing our Stellar Gula near Hercules. In *Mul.Apin* I i 24–25 the constellation before "The Dog" in the list of Enlil-stars is "The She-Goat" (^{mul}ÙZ), who is there associated with the goddess Gula. In several texts that make use of the *ziqpu*-stars, "The She-Goat" appears in place of the more common

[39] Kurtik (2007, 119–22), sub. ^{mul}DINGIR.GUB.BA^{meš} and ^{mul}DINGIR.KU.A^{meš}.

[40] Nougayrol (1947, 31).

ziqpu-star "The Lady of Life" (^{mul}GAŠAN.TIN), a not-inappropriate appellation for the healing goddess Gula.[41] Thus, we suggest that "Stellar Gula" is a constellation that includes part or all of "The She-Goat" and takes the place of that constellation in the repertoire of Enlil-stars used in this text. The placement of "Stellar Gula" after "The Dog" in our text, whereas "The She-Goat" appears before "The Dog" in *MulApin*, would seem to indicate that the asterism Stellar Gula consisted of a complex picture of multiple constellations extending to both sides of "The Dog."

A second goddess Gula in the sky, albeit without the MUL-sign, is to be found in the vicinity of "The Twins" (also written without the MUL-sign) and "The Crab" in the Alb B star-catalogue:

> ¶ MUL SA_5 *ša ina* ZI IM.KUR.RA EGIR ^{d}MAŠ.TAB.BA DA ^{d}GU.LA GUB-*zu* AGA *ap-ru* ^{mul}*Al-lu-ut-tum* MUL ^{d}*A-nim* LUGAL
>
> (Alb B II ii rev. 5–8)
>
> The red star which stands at the rising of the eastwind after "The Twins" alongside Gula, wearing a crown, "The Crab," The star of Anu, the King.

This second Gula cannot be the same constellation as that described in our text due to the distance between "The Dog" on one hand and "The Twins" and "The Crab" on the other.

A rev. 16: We restore the number of stars, two, at the start of the line on the basis of A rev. 7. Akkadian *kablu* refers to the legs of pieces of furniture, here a chair, but elsewhere also beds, tables, and potstands. The presence of the chair indicates that the goddess depicted by the constellation is seated, so confirming the imagery of Sections IX–X, the goddess Gula seated with her dog. We imagine that such a drawing of the two constellations together was known to the author.

Section XI–XII: A rev. 17–21

Sections XI–XII on the bottom edge of A are too broken to be identified with any particular star, and do not match anything in D that occurs after the material that parallels Section VII at the end of D iii.

Rev. 21: For 2 stars "side by side" (Á *ana* Á), see previously A rev. 7 and 16.

41 Roughton et al. (2004, 549); Steele (2014, fn 37).

CHAPTER 3

The Expanded Version

Edition

D. MLC 1866. Photographs: plates 8–9; Copy: plates 10–14.

D. Upper Edge

⸢*ina*⸣ [*a*]-⸢*mat* ^{d}60 *u an-tu*$_4$ *liš-lim*⸣

OBVERSE

D col. i

Section A

D i 1. [¶] ⸢^{mul}IKU⸣ *šu-bat* ^{d}60 *a-lik pa-na-at* ⸢MUL.MEŠ *šu-ut* ^{d}60⸣

D i 2. ⸢^{itu}⸣ [BAR]Á ZAG.MUK *šu-bat kip-pat*^{meš} *er-bet-tu*$_4$ MUL *reš*[1]

D i 3. [4 MUL] ⸢Á⸣ *ana* Á *e-ṣir-u'* : *šat-ti*

D i 4. [MUL *šá ina*] *mih-rat* ^{mul}IKU GUB-⸢*zu*⸣ ^{mul}*Sí*-⸢*nu*⸣-*nu-tú*

D i 5. ⸢MUŠEN⸣ MUL KAP.HI.A *mut-tap-ri-iš šá kap-pi* ⸢*ra*⸣[2]-*šu-ú*

D i 6. MUL *šá* EGIR.MEŠ ^{mul}IKU GUB-*zu* ^{mul}*A-nun-ni-tu*$_4$ *na-a-ru*

D i 7. [^{mu}]^{l}*Sí-nu-nu-tu*$_4$ *ù* ^{mul}*A-nun-ni-tu*$_4$ *ina* KUN.⸢MEŠ-*šú*⸣-*nu it-gu-ru-ú-ma*

D i 8. ⸢MUL⸣.MEŠ ^{íd}IDIGNA *u* ^{íd}BURANUN^{ki} *ina ṣip-ri* ^{mul}ZIB

D i 9. ⸢*ṣab-tu*$_4$⸣ ^{mul}SIM.MAH *u* GÚ.LÁ ^{mul}*A-nun-ni-tu*$_4$ *man-za-za* ^{mul}DILI.BAD

D i 10. ^{mul}IKU *a-na* AB.SÍN *u* ^{mul}IKU *ina* ^{itu}BAR IGI BURU$_{14}$ KUR[3]

D i 11. ^{mul}SAG.ME.GAR ^{mul}DILI.BAD *u* ^{mul}GU$_4$.UTU KI-*šú* : SI.SÁ

D i 12. [*in-n*]*a-mir* BURU$_{14}$ KUR SI.SÁ : MUL *šá* EGIR-*šú* GUB-*zu*

D i 13. ^{mul}LÚ.LU.HUN.GÁ ^{mul}UDU/LU.HUN.GÁ ^{itu}BAR *ár-hu* ^{d}60

D i 14. MUL *re-eš šat-ti šu-ú* ^{mul}LÚ.HUN.GÁ *as-lum im-mer*

D i 15. ^{d}*Dumu-zi šá* ^{mul}LU/UDU.HUN.GÁ 3 MUL *ina* SAG.KI-*šú e-ṣir-u'*

D i 16. ⸢2⸣ MUL *ina* GIŠ.KUN-*šú e-ṣir-u'* : 4 ⸢MUL⸣ *ina* GÌR.MEŠ-*šú* GUB-*zu-u*[4]

---

[1] The scribe ran out of space at the end of the line and so wrote the next word *šat-ti* at the end of the next line following the *Glossenkeil*. Cf. D i 10–11. This use of the *Glossenkeil* suggests that our scribe was following the arrangement of the text on a previous source.

[2] The RA sign is very small and squeezed in between PI and ŠU. It seems to have been inserted as an afterthought after being accidentally omitted.

[3] The last word of this sentence is given at the end of line 11 following the *Glossenkeil*. Cf. lines D i 2–3. The subject of the main portion of line 11 then continues on through the middle of line 12 where it is marked by a second *Glossenkeil*, which is followed by the start of a new chain of thought concerned with "The Hired Man."

[4] The purpose of the *Glossenkeil* in this line is not clear. Perhaps it is intended to indicate that the two statements on the current line D i 16 were written on separate lines on a previous source. This may also be the case for the *Glossenkeil* in D i 17.

D i 17. ⸢¶⸣ ^{mul}UD.KA.DUH.A ^{d}U.GUR : SAG.KAL *a-lik* ⸢*mah-ri*⸣[5]
D i 18. ⸢*ṣa-lam*⸣ *šú-u lu-bu-uš-tu*$_4$ *la-biš* AGA a-pir
D i 19. ⸢2⸣ *pa-ni-šú pa-ni mah-*⸢*ru*⸣*-tú pa-ni ṣa-lam*
D i 20. ⸢*pa*⸣*-ni ár-ku-*⸢*tu*$_4$ *pa*⸣*-ni* ⸢UR⸣.MAH *ziq-nu za-qin*
D i 21. *pi-i pi-*⸢*ti kap*⸣*-pi ra-áš-ši kin-ṣi* 15*-šú*
D i 22. ⸢*ki-ṣal*⸣*-šú* 3.⸢TA⸣.ÀM MUL *a-si-du* 15*-šú* KI.ÚR
D i 23. [x x x *a-si-d*]*u* KI.⸢ÚR⸣ *ṣa-lam*
D i 24. [MUL *šá ina* ZAG-*šú* GUB]-⸢*zu*⸣ ^{mul}ŠAH ⸢MUL⸣ *šá ina* GÙB-*šú*
D i 25. [GUB-*zu* ^{mul}ANŠE.KUR.RA] x x *a-n*[*a*] x x x
D i 26. [ ] x x [ ]

Rest of Column i lost

D col. ii

D ii 1. *hi-*⸢*in*⸣*-šu* : *ku-ut-ta-ú ina* ŠU.MIN 15*-šú na-áš-ši* [*ina* ŠU.MIN] ⸢2⸣,30*-šú*
D ii 2. *il-tuh-hu na-áš bir-qu it-ti il-tuh-hu ṣa-bit*
D ii 3. *ṣa-lam* EGIR*-ú ina* ŠU.MIN 15*-šú pa-a-šú na-áš ina* ŠU.MIN 2,30-*šú il-*⸢*tuh*⸣*-hu*

D ii 4. ¶ ^{mul}MAŠ.TAB.BA.TUR.TUR ^{d}MAŠ.MAŠ TUR.MEŠ ^{d}*tu-ma-mu*^{meš} TUR.MEŠ
D ii 5. *šá ina pa-na-at* ^{d}60 GUB-*zu-u'* 2 *ṣa-lam lu-bu-uš-tu4 lab-šu-u'*
D ii 6. *ziq-nu ziq-na-a' ku-ur-ku-ru šak-na-a' ina* SAG.MEŠ-*šú-nu* 2
D ii 7. MUL.MEŠ *e-ṣir-u' ṣa-lam* IGI*-ú ina* ŠU.MIN 15*-šú il-tuh-hu na-áš*
D ii 8. *dir-rat it-ti il-tuh-hu ṣa-bit ṣa-lam* EGIR*-ú ina* ŠU.MIN
D ii 9. 15*-šú bir-qu na-áš ár-ku* ^{mul}SIPA.ZI.AN.NA GUB-zu

---

D ii 10. ¶ ^{mul}SIPA.ZI.AN.NA ^{d}*pap-sukkal suk-kal-lum* ^{d}60 *u an-tu*$_4$
D ii 11. *ṣa-lam šu-ú lu-bu-uš-tu*$_4$ *la-biš ziq-nu za-qin ku-ur-ku-ru šá-*⸢*kin*⸣
D ii 12 MUD *u nam-za-qa ta-mi-ih* ^{mul}MAŠ.TAB.BA *šá ina* ⸢*pa*⸣*-ni*
D ii 13. ⸢^{mul}⸣SIPA.ZI.AN.NA GUB-*zu* ^{d}LÚ.LÀL *u* ^{d}⸢*la-ta*⸣*-ra-ak-a*
D ii 14.[6] *šá* ⸢KÁ⸣*-a-ni* 2 *ṣa-lam lu-bu-*⸢*uš*⸣*-t*[*u*$_4$ *l*]*ab-šu-u' ṣa-lam* IGI*-ú*
D ii 15. *ziq-nu za-qin ṣa-*[*lam ár*]*-ku-ú pa-n*[*i* ^{d}]*la-ta-ra-ak-a*
D ii 16. *ku-ut-ta-ú ina* ŠU.MIN 15*-šú-nu na-šu-ú* ⸢MUL⸣ *šá ina šap-li*
D ii 17. ^{mul}SIPA.ZI.AN.NA GUB-*zu* ^{mul}DAR.LUGAL

D ii 18. ¶ ^{mul}ALLA *šu-bat* ^{d}60 *áp-sà-*⸢*am-ma*⸣*-ku* 4 MUL *ina* Á.MEŠ*-šú*
D ii 19. *e-ṣir-u' ina lìb-bi-šú* 6 MUL *a-hi-nu-ú a-*[*h*]*a-meš* U$_5$*-u'*
D ii 20. ⸢1+*en*⸣ MUL *ina* SAG*-šú e-ṣir ina pa-n*[*i-šú* ^{mul}SAG.ME.GAR GU]B-*zu*

---

[5] The *Glossenkeil* here may serve the same purpose as that in the previous line; see fn 4. Another possibility is that it is meant to separate the quote from *Mul.Apin* in the first half of the line from what comes next.
[6] D i 15–16 intrude, pushing the start of this and the next line slightly to the right.

D ii 21. [¶] ^{mul}KAK.PAN ^{d}IM DUMU ^{d}60 *u* [*an-tu*$_4$ GÚ.GAL AN-*e*]
D ii 22. *ù* KI-*tì mih-rat* ^{mul}ALLA [x x x x x]

D ii 23. [¶ ^{m}]^{ul}PAN ^{d}*ša-la ru-ba-at* [x x x]
D ii 24. *ana ṣer-ret* [AN-*e* . . .]

Rest of Column ii lost

D col. iii

D iii 1. ¶ ^{mul}M[UŠ] x x x x x x *šú-ú*
D iii 2. *kap-pi šá-kin* GÌR.MIN *ra-áš* UGA^{mušen} MUL ^{d}IM
D iii 3. *ina muh-hi* KUN-*šú*

D iii 4. ¶ ^{mul}*Nin-mah* ^{d}GAŠAN DINGIR.MEŠ *pu-uš-ku* A.ME
D iii 5. *lu-bu-uš-tu*$_4$ *lab-šá-at ku-ur-ku-ru šak-na-at*
D iii 6. TIR-*ri ina muh-hi ina* ŠU.MIN 15-*šú ù* 2,30-*šú* ⸢*na-šá-at*⸣

D iii 7. ¶ ^{mul}EN.TE.NA.BAR.HUM ⸢^{d}*Nin*⸣-[*gír-s*]*u*
D iii 8. *ṣa-lam* SAG.DU 1+EN M[UL x x x x]
D iii 9. *ina* ⸢ŠU.MIN⸣ [15$^{?}$]-⸢*šú*⸣ x x [x x x x x x x]
D iii 10. x [. . . ¶ 2 MUL.MEŠ]
D iii 11. ⸢*šá* EGIR-*šú*⸣ GUB-*zu* ^{d}PA ⸢^{d}⸣[LUGAL ^{d}UTU ^{d}IM][7]
D iii 12. MUL *šá* EGIR GUB-[*zu* ^{mul}*Nu-muš-da* ^{d}IM]

D iii 13. ¶ ^{mul}MAR.GÍD.DA *e-re-eq-*[*qu* ^{d}*Nin-lil*]
D iii 14. GIŠ.GIGIR ^{d}*a-nu-um áp-sà-am-*[*ma-ku* 4 MUL.MEŠ]
D iii 15. *ina* Á.MEŠ-*šú e-ṣir-u'* 3 MU[L.MEŠ *ina* SAG.KI-*šú*]
D iii 16. *e-ṣir-u'* ^{mul}GÀM *ina* x-[x-*šu e-ṣir*]
D iii 17. 2 MUL.MEŠ *un-nu-tu-tu* Á *ana* ⸢Á⸣ [IGI-*šú ina ma-šad-di*][8]
D iii 18. *e-ṣir-u'* MUL *šá it-ti za-r*[*u* ^{mul}MAR.GÍD.DA]
D iii 19. GUB-*zu* ^{mul}⸢KA$_5$.A ^{d}*Ér-ra*⸣ *gaš*$^{!}$-*ri*[9] D[INGIR.MEŠ]
D iii 20. MUL *šá ina* SAG.KI ^{mul}MAR.GÍD.DA DIŠ x [. . .]
D iii 21. ^{mul}*im-mer-tu*$_4$ ^{d}⸢*e*$^{?}$⸣-*a* MUL⸣ *š*[*á* . . .]
D iii 22. GUB-*zu an-tu*$_4$ GAL-*tu*$_4$ *šá* A[N-*e* . . .
D iii 23. [x x G]UB-⸢zu⸣ x [. . .

Rest of column lost (substantial break)[10]

---

[7] D iii 11–12 are restored on the basis of *MulApin* I ii 25–27. See the commentary.
[8] Restored on the basis of ABC Section VII.
[9] BE-*ri*, instead of BI(*gaš*)-*ri*.
[10] The lower half of col. iii on the obverse and upper half of col. iv on the reverse.

REVERSE

D col. iv

Beginning of column lost

D iv 1'. x [. . .

D iv 2'. DIŠ M[UL$^?$ [. . .

D iv 3'. ¶ ^{mu}[^{l} . . .

D iv 4'. *lu-bu-*[*uš-tu*$_4$ *la-biš* . . .

D iv 5'. *ku-*⸢*ur-ku*⸣-[*ru šá-kin* . . .

D iv 6'. *qup-pe-e* [x x] ⸢LU⸣ [. . .

D iv 7'. ¶ ^{mul}KU$_6$ ⸢^{d}⸣*É-a* x [x x x x]

D iv 8'. ¶ ^{mul}SAG.ME.GAR ^{mul}*dil-bat* ^{mul}G[U$_4$.UTU ^{mul}GENN]A

D iv 9'. ¶ ^{mul}UDU.IDIM ^{mul}*ṣal-bat-an-nu* ^{d}30 *u* ^{d}UTU

D iv 10'. 7 DINGIR.MEŠ DUMU.MEŠ ^{d}*a-num šá ina ri-*⸢*ha*⸣*-tú* ^{d}60

D iv 11'. *re-hu-ú* ^{d}*i-gi*$_4$*-gi*$_4$ ^{d}⸢*pap-sukkal*⸣

D iv 12'. *ma-lik* ^{d}60 *si-bit-ti-šú-nu ina šu-bat* ^{d}60

D iv 13'. LUGAL *rab-biš* G[UB.Z]U-⸢*u*'⸣*-ma* GABA.RI

D iv 14'. *ul i-šu-*⸢*ú*⸣

D iv 15'. ¶ ^{mul} IBILA.<É>.MAH ^{d}*É-a* ⸢MUL⸣ [DU]MU É

D iv 16'. ^{mul}MAŠ.TAB.BA.GAL.GAL ^{d}LUGAL.ÌR.RA

D iv 17'. *ù* ^{d}MES.LAM.TA.È ⸢*suk-kal-lum*⸣ ^{d}60

D iv 18'. ^{mul}MAŠ.TAB.BA.TUR.TUR ^{d}MAŠ.MAŠ ^{d}*tu-ma-mu*^{meš}

D iv 19'. *šá ina pa-na-at* GUB-*zu* SUKKAL ^{mul}SIPA.ZI.AN.NA

D iv 20'. ^{d}*pap-sukkal* SUKKAL ^{d}60 *u an-tu*$_4$

Col. v (colophon and date)

Beginning of column lost (substantial break)[11]

1'. DIŠ SI[G$_7$ . . . .]

2'. DIŠ DINGIR.GAL.GAL.⸢E⸣.N[E x x x x]

3'. ⸢GIN$_7$⸣ SUMUN-bar-*šú* ⸢SAR$^?$*-ma*⸣ IGI *ù u*[*p*$^?$*-puš*$_4$ ŠU.MIN . . .]

4'. DUMU *šá* ^{Id}60.PAP.GÁL-*ši* A *šá* ^{m}*é-kur-za-kir* LÚ MAŠ.MAŠ ^{d}*a-nu-um*

5'. *ù an-tu*$_4$ LÚ ŠEŠ.GU.LA *šá* É *re-eš* ⸢LÚ⸣ *šá-as-suk-ku*

6'. *šá* ⸢U$_4$ AN⸣ ^{d}*en-líl-lá šá ina pi-qí-*⸢*ti* ^{m}⸣*a-da-pu*

7'. ⸢^{lú}*um*⸣-[*m*]*án*-ME ME KI ITI ⸢GAN U$_4$⸣ 23.KAM *šat-ti* 97.KAM

8'. ^{m}*an-t*[*i-'u-ku-u*]*s-su* LUGAL KUR.KUR

---

[11] The use of single and double spacing between lines in the transliteration and translation of the colophon follows the format of the original tablet where the scribe leaves more than usual space between some lines.

Col. vi[12]

1'. . . . .] x
2'. . . . .] x x
3'. [x x x] x x x x x x x x
4'. [x x x] x x x x x x x x
5'. [x x] x x x x x x x x [x] x x
6'. [x] x x x x x x x x [x] x x
7'. [x x] x x x x x x x x x x

Translation

Upper Edge

By the [co]mmand of Anu and Antu, may this go well.

OBVERSE

D col. i

Section A

i 1. ¶ "The Field," the seat of the god Anu, the forerunner of the stars of Anu,
i 2. the month of [Nisa]n, the New Year's Festival, the seat of the circle of the four quarters, the star of the New Year.
i 3. [4 stars] are drawn side by side.
i 4. [The star which] stands opposite "The Field," (is) "The Swallow":
i 5. (It is) a bird, a star with wings, flying, that is it has wings.
i 6. The star which stands after "The Field" is Anunitu, a river.
i 7. ["Th]e Swallow" and Anunitu at their tails cross one another,
i 8. the stars of the Tigris and Euphrates, by the crest of "The Tail"
i 9. they are held together. "The Swallow" and the stretched out neck of Anunitu, the station of Venus.
i 10. "The Field" concerns the furrow and "The Field" in Nisan rises heliacally – the harvest of the land will prosper,
i 11. Jupiter, Venus or Mercury with it
i 12. [is s]een – the harvest of the land will prosper: The star which stands after it is
i 13. "The Hired Sheep Man" (Aries = ^{mul}LÚ.HUN.GÁ), "The Hired Sheep,"[13] Nisan, the month of Anu,
i 14. it is the star of the New Year. "The Hired Man" (Aries), the lamb, ram,
i 15. Dumuzi. Concerning "The Hired Sheep" – three stars are drawn at its forehead;
i 16. two stars are drawn on its thigh: four stars stand at its feet.

Section B

i 17. ¶ "The Demon with The Open Mouth," Nergal: the leader going in front,
i 18. He is a human figure, clothed, wearing a tiara.
i 19. Two are his faces. The front face is the face of a human figure.

---

[12] Col. vi is too worn for transliteration.
[13] The translation attempts to convey the sign-play in the original; see the commentary.

i 20. The back face is that of a lion, bearded,
i 21. mouth open, having wings; its right shin (and)
i 22. its ankle are three fold; the star of its right heel (and) *lower extremities*
i 23. [(are) . . . the hee]l (and) the *lower extremities* of a human figure.
i 24. [The star which at its right stand]s is "The Pig"; the star which at its left
i 25. [stands is "The Horse,"] . . fo[r] . . .[14]
i 26. [ ] . . [ ]

D col. ii

Section C

ii 1. (The front human figure of "The Great Twins") carries a *hinšu* : *a large jug*? in his right hand; [in] his left [hand]
ii 2. he carries a whip; he holds a lightning bolt with the whip.
ii 3. The back human figure carries a sickle-axe in his right hand, in his left hand, a whip

---

Section D

ii 4. ¶ "The Lesser Twins," "The Small Twins," "The Small Gemini,"
ii 5. who stand in front of Anu. Two clothed human figures,
ii 6. bearded, set with a *kurkurru*; at their heads two
ii 7. stars are drawn. The front human figure carries a whip in his right hand,
ii 8. holds a lash with the whip. The back human figure carries a
ii 9. lightning bolt in his right hand, stands behind "The True Shepherd of Anu."

Section E

ii 10. ¶ "The True Shepherd of Anu," Papsukkal, the vizier of Anu and Antu.
ii 11. He is a human figure, clothed, bearded, set with a *kurkurru*,
ii 12. grasping a lock and key. "The Twins" which in front of
ii 13. "The True Shepherd of Anu" stand, Lulal and Latarak
ii 14. *of the gates*.[15] Two human figures, cl[ot]hed. The front human figure
ii 15. is bearded. The [ba]ck human fig[ure] has the fac[e] of Latarak;
ii 16. they carry *a large jug* in their right hands. The star which below
ii 17. "The True Shepherd of Anu" stands is "The Rooster."

Section F

ii 18. ¶ "The Crab," the seat of Anu, an *apsamakku*. Four stars at its sides
ii 19. are drawn; inside it six stars side by side straddle o[n]e another,
ii 20. one star is drawn at its head, in fro[nt of it Jupiter stan]ds.

---

[14] For a possible restoration of the last half of the line, see the commentary.
[15] Either "gates" (*bābu*, plural) or CAD B 7 *bābānu*, "outside"; see the commentary.

Section G

ii 21. [¶] "The Arrow," Adad, the son of Anu and [Antu, The Canal Inspector of Heaven]
ii 22. and Earth, opposite "The Crab" [. . . . .].

Section H

ii 23. [¶: "T]he Bow," Šala, the princess [. . .]
ii 24. for the lead-rope(s)/breasts [of heaven . . .]

Rest of column lost

D col. iii

Section I

iii 1. ¶ "The S[nake"] . . . . . . it is,
iii 2. equipped with wings, having feet; a raven, the star of Adad
iii 3. (is) on top of its tail.

Section J

iii 4. ¶ Ninmah, the mistress of the gods, *a pušku? of water*.
iii 5. She is clothed, set with a *kurkurru*,
iii 6. carries a . . in front in her right and left hands.

Section K

iii 7. ¶ EN.TE.NA.BAR.HUM, Nin[girs]u,
iii 8. a human figure, the head is one s[tar . . . .]
iii 9. in its [right?] hand . . [. . . . . .]
iii 10. . [. . . The two stars]
iii 11. which stand after it are Šullat (and) [Haniš, Shamash (and) Adad.]
iii 12. The star which stand[s] after [is Numušda, Adad]

Section L

iii 13. ¶ "The Wagon," a wago[n (Akkadian), Ninlil]
iii 14. the chariot of Anu. An *apsam*[*akku*, 4 stars]
iii 15. at its sides are drawn; three star[s at its fore]
iii 16. are drawn. "The Crook" at [its] . [. is drawn].
iii 17. Two faint stars, side by side [in front of it on the pole]
iii 18. are drawn. The star which with the cart-po[le of "The Wagon"]
iii 19. stands, "The Fox," Erra, the strong one among the g[ods].
iii 20. The star which in front of "The Wagon" . . [. . .]
iii 21. "The Ewe," Aya.[16] The star wh[ich . . .]

---

[16] For the reading of Aya, see the commentary.

iii 22. stands, Great Antu of Hea[ven . . .
iii 23. [. . st]ands . [. . .

Rest of column lost

REVERSE

D col. iv

Beginning of column lost

Section M

iv 1'. . [. . . . .
iv 2'. 1 s[tar$^{?}$ . . . .

Section N

iv 3'. ¶ "The (sta[r") . . . .
iv 4'. clo[thed . . . .
iv 5'. [set with] a *kurku[rru* . . .
iv 6'. a knife [. .] . [. . .

Section O

iv 7'. ¶ "The Fish, Ea, . [. . . . .]
iv 8'. ¶ Jupiter, Venus, Me[rcury, Satur]n;
iv 9'. ¶ The planet Mars, the Moon and the Sun.
iv 10'. Seven gods, sons of Anu, who by the seed of Anu
iv 11'. were begotten, the Igigi-gods. Papsukkal (is)
iv 12'. the counselor of Anu. The seven of them in the abode of Anu,
iv 13'. the king, magnificently they s[tan]d, a rival
iv 14'. they have not.

Section P

iv 15'. ¶ "The Heir of the Sublime <Temple>," Ea, *The Star*, [*so*]*n of the temple.*[17]
iv 16'. "The Great Twins," Lugalirra
iv 17'. and Meslamtae, the vizier of Anu.
iv 18'. "The Lesser Twins," the Twins, the Gemini
iv 19'. who stand in front. The vizier (is) "The True Shepherd of Anu,"
iv 20'. Papsukkal, the vizier of Anu and Antu.

D col. v (colophon and date)

Beginning of column lost

v 1'. . [. . . . .]

v 2'. . the great god[s . . . .]

---

[17] Akkadian *mār bīti*.

v 3'. According to its original, *written*?, checked and ex[ecuted, at the hands of...],

v 4'. son of Anu-ah-ušabši, descendant of Ekur-zākir, the exorcist of Anu

v 5'. and Antu, high priest of Bit Reš, recorder

v 6'. of (the series) *Enuma Anu Enlil*, which is under the authority of Adapa,

v 7'. the Sa[g]e of...:[18] The month of *Kislīmu*, 23rd day, Year 97 (of the Seleucid Era)

v 8'. Ant[ioch]us, king of the lands.

D col. vi

vi 1'–7'. Traces, surface too worn to allow for translation

Commentary

Section A: D i 1–16, ^{mul}IKU = *ikû*, "The Field"

"The Field" appears in the list of stars in the path of Anu in *Mul.Apin*. In the so-called TE tablet (BM 77824), the entry for Month XII gives "The Field" alongside "The Tails." In the late period, "The Field" is occasionally used an alternative to "The Tails" as the name of the zodiacal sign Pisces.

D i 1: This line quotes *Mul.Apin* I i 40, but with the star as "the seat of Anu" rather than Ea.[19] This discrepancy is without doubt due to the influence of the Anu-cult at Uruk. In D, as in *Mul.Apin*, "The Field" is the forerunner of the stars of Anu. In Alb B ii 1 the constellation is the forerunner of the stars of Ea. Note in D the syllabic writing *a-lik pa-na-at* rather than *a-lik* IGI as in *Mul.Apin* and the Astrolabes.

D i 2: For "The Field" identified with the city of Babylon, the month of Nisan, and so the New Year at Babylon, see previous discussion in George (1992, 244–5) and Horowitz Alb (56). The term *kippat erbetti*, "the circle of the four quarters," refers to the traditional division of the Earth's surface into north, south, east, and west quadrants in ancient Mesopotamian thought, with the circle referring to the perceived shape of the Earth's surface, or at least the central continental area of the Earth's surface, as on the *Babylonian Map of the World*.[20] Thus, the identification of "The Field" as *šubat kippat erbetti*, "abode of the circle of the four quarters," refers to the constellation as Babylon in its guise as world and cosmic capital city. For a drawing of a circle with wind directions indicated which might have illustrated the concept of *kippat erbetti* when complete, see BagM Beih 2, no. 98.[21]

---

[18] For the three untranslated signs ME ME KI, see the commentary.

[19] The writing ^{d}DIŠ could in theory be for either Anu (^{d}60) or Ea (^{d}1), but ^{d}60 is the standard writing for Anu's name in our text, whereas Ea's name is written syllabically (D iv 8, 15). See, e.g., the upper edge of D for Anu (^{d}60) and Antum.

[20] For *The Babylonian Map of The World*, see Horowitz (2011, 20–42). For the four regions of the Earth's surface, and the various terms used to express this idea, see ibid., 298–9, 334.

[21] Horowitz (2011, 193–207).

D i 3: The restoration [4 stars] fits the protases of BPO 2 64 Text XII 1–4 which speak of northern and southern, and eastern and western, stars in the constellation "The Field"; that is, what would seem to be here two sets of two stars placed "side by side."[22] These four stars may be presumed to form the rectangle consisting of the stars α, β, and γ Pegasi, and Sirrah (α Andromedae), now known as the "square of Pegasus," thus creating the square field that gave the star its ancient Mesopotamian name. The four stars of "The Field" may also echo the constellation's identification with Babylon as "abode of the circle of the four quarters" in line 2.

D i 4: This line quotes *MulApin* I i 41. "The Swallow" and the next star in D, Anunitu, comprise the classical constellation Pisces. See the commentary to D i 7.

D i 5: The traces of the nearly effaced first sign of the line are consistent with the sign HU for MUŠEN (*iṣṣūru*), "(it is) a bird . . .," which describes "The Swallow" named at the end of D i 4, and with the description of the constellation that follows. We take KAB here as a sort of pseudo-sumerogram for KAB = *kappu*, "wing."[23] This equation is attested much earlier than our text in Proto-Aa 464 1 (MSL 14 100). Cf. MSL 14 126 807.

D i 6: This line quotes *MulApin* I i 42, but with EGIR.MEŠ for *arkāt* paralleling *pa-na-at* in col. i 1, ii 5, and iv 19'. The reference to Anunitu as a river is explained in lines 7–8 where "The Swallow" and Anunitu are identified with the Tigris and Euphrates.[24]

D i 7–9: The constellations "The Swallow" and Anunitu, with their tails crossed, form the classical constellation Pisces, which was known in first millennium Mesopotamia under the name ^{mul}ZIB/ZIB.ME = *zibbāti*, "The Tails."[25] This configuration was connected with fish in ancient Mesopotamia as well.[26] Note for example the reference to two fish together in BPO 3 248 48–49, and the name of the Normal Star in classical Pisces, MÚL KUR *šá* DUR *nu-nu*, "The Bright Star in the Ribbon of the Fishes," which suggests at least two fish held together by string.[27] In any case, an image of "The Swallow" and Anunitu as two fish joined by crossed tails was not the only one that circulated. In our lines 8–9, "The Swallow" and Anunitu are held together by the *ṣipru* of a tail-star (^{mul}ZIB). Akkadian *ṣipru* bears allusions to birds but is also to be found in astronomical contexts (CAD Ṣ 204–205). This brings to mind the identification of "The Swallow" constellation as a winged bird in D i 5, suggesting a Mesopotamian view of classical Pisces as a bird ("The Swallow") and a fish (Anunitu) tied together by a cord. This matches drawings of what have been

---

22 For these four stars, see Kurtik (2007, 79–80).

23 KAB = *kappu* can be compared with ZIB = *zibbātu*, "tails," in D i 8.

24 For the identification of "The Swallow" (the western fish of Pisces) with the Tigris, and Anunitum (the eastern fish of Pisces) with the Euphrates, see Weidner (1963), Horowitz Alb (113–4), and Kurtik (2007, 436). The identification of the two parts of classical Pisces with the two rivers may relate in some way to the shape formed by "The Swallow" and Anunitum, which can be visualized as two lines that come together at their end, like the Tigris and Euphrates themselves which join to form the Shatt-al-Arab, whose ancient equivalent was known to the Babylonians as the *bitqu*, "sluice" (Horowitz 2011, 28–9).

25 See Kurtik (2007, 599–601), CAD Z 102 *zibbatu* 3.

26 This was in addition to the traditional Sumero-Akkadian fish-constellation, ^{mul}KU$_6$ = *nūnu*, "The Fish" = Piscis Austrinus.

27 See also Kurtik (2007, 66); Roughton et al. (2004, 566) (DUR sim-mah and DUR ^{múl}*nu-nu*, both of which occur in BM 36609+), and CAD R 348 *riksu* 1d. However, *nu-nu*, as written, could also be interpreted as singular, for only one fish, as seems to be the case for Anunitu as a fish in Urra XXII 303: ^{mul}nu-nu = ^{d}*a-nu-ni-tum*; *nu-nu* here in the Sumerian column is a loan-word from the Akkadian word for fish, *nūnu*.

taken to be "The Swallow" and Anunitu as a bird and fish, with their tails tied together; see Kurtik (2007, 738–9, no. 45–46), and cf. no. 44 (with their tails not attached).[28] The bond holding the two parts of classical Pisces together is what must be meant by ^{mul}ZIB at the end of line D i 8, and it is also described in *Aratus: Phaenomena* (Kidd 1997, 239–47):

> Still farther ahead and more in the approaches to the south are the Fishes; but one is always more prominent than the other and hears more the fresh onset of the north wind. From both of them stretch, as it were, chains from the tail-parts, coming together in an unbroken line on both sides. One beautiful bright star occupies this position, and they call it the celestial Knot (ibid., 91)

Aratus' "celestial Knot" is to be identified with the star Al-Risha, which in Arabic means "The Cord," a name that probably harkens back to earlier Mesopotamian views of a bond joining "The Swallow" and Anunitum together.

D i 9: Our translation here is literal: GÚ = *kišādu*, "neck," and LÁ = *ṭarāṣu*, "to extend/stretch out a part of the body."[29] In the aforementioned drawing of classical Pisces (Kurtik 2007, 738, no. 45), it is the bird, presumably "The Swallow," that has a stretched-out neck, not Anunitum as in D i 9. The reference to the "station of Venus" must refer to the planet's *bīt niṣirti*, which is located in the region of Pisces. This suggests that if the Pisces portion of the micro-zodiac series published by Weidner (1967) was preserved, Venus would be drawn along with the Swallow and Anunitum, in the same way as Jupiter, Mercury, and the moon appear on drawings for Leo, Virgo, and Taurus, respectively.

D i 10–12: Lines 10–11, and the first half of line 12 until the *Glossenkeil*, continue the discussion of "The Field" which began at the start of Section A. What we seem to have here are two notices concerning the constellation, joined by the conjunction "and." The first is at the very start of line 10, offering an association between the constellation "The Field" and the furrow that is formed by means of the preposition *ana*.[30] Such associations formed by *ana* are common in Mesopotamian astrology, with many to be found in Assumed EAE 51 in BPO 2 and the Astrolabe type star-catalogues BM 55502//HS 1897.[31] The second notice consists of what is written as four direct statements, with parts 1 and 3 relating to the "The Field" and parts 2 and 4 relating to the harvest. However, when read consecutively, the four statements read like two astronomical omens in which the protasis relates to "The Field" rising in the first month of Nisan, and the apodosis to the harvest,[32] with all this presumably reflecting the role of "The Field" as the star of the New Year that is noted in D i 2.

D i 12–15: The entry for "The Hired Man" begins like the start of the star's entry in *MulApin* I i 43: ¶ MUL *šá* EGIR-*šú* GUB-*zu* $^{mul\ lú}$HUN.GÁ ^{d}*Dumu-zi*, "The star which stands after it is 'The Hired Man,' Dumuzi."

[28] See also further discussion in Wallenfels (1993, 287) and White (2007, 47–8). Hallo (2008, 239) points out that the combination of birds and fish as a symbol of abundance is already attested in Sumerian times.

[29] CAD T 208, albeit not attested in the dictionary with the neck.

[30] In the absence of a star-determinative with AB.SÍN, we do not take this to be for the constellation "The Furrow" (^{mul}AB.SÍN).

[31] BPO 2 58–59, Text IX 16–34, 64–65, Text XII, with discussion and further examples available on BPO 2 25 and Horowitz Alb (109–10). For HS1897//BM55502, see Horowitz Alb (101–6).

[32] 1) (If) "The Field" in Nisan rises heliacally, the harvest of the land will prosper.
2) (If) Jupiter, Venus, and Mercury with it are seen, the harvest of the land will prosper.

In Mesopotomia, the constellation ^{mul}lú-hun-gá = *agru* is a human figure, but its later classical equivalent is Aries, "The Ram." This transformation from human form to sheep figure is observed by the authors of our text. In line 12, the name of the star is written out twice with most unorthodox orthographies. Instead of ^{mul}LÚ. HUN.GÁ, we first find ^{mul}LÚ.LU.HUN.GÁ, and then ^{mul}LU/UDU.HUN.GÁ.[33] In the former, LU appears alongside the expected homonym LÚ; the sign for "human being/man." In the latter, LÚ is dropped altogether leaving only the homonym LU. This is then explained at the end of line 14 where the constellation is said to be a lamb and ram, justifying the use of the LU-sign, apparently to be understood with its common Sumerian reading and meaning udu = *immeru*, sheep as a general category.[34] Thus, what we have here is a most clever sign-play, sign-exegesis if you like, where the cuneiform name of the constellation moves from human (LÚ) form to sheep (LU/UDU) form by means of the homonym, just as the constellation itself was transformed in late Mesopotamia from the older human form, as "The Hired Man," to the sheep form of classical Aries, "The Ram."[35] This play of signs also lies behind the abbreviation of LU = udu for Aries attested elsewhere in late Mesopotamia, particularly in astronomical texts (especially those containing mathematical astronomy) from Seleucid period Uruk.[36] At Babylon, Aries is usually written HUN. This may suggest that the association of Aries with a "Ram" was an Uruk invention and remained predominantly an Urukean tradition. It is possible, but by no means necessary, that the transformation of Aries into a "Ram" reflects Greek influence.[37]

D i 13: The identification of the constellation with the New Year in lines 13–14 can be explained on the basis of *Mul.Apin* I ii 36, where New Year's Day, the first of Nisan, is given as the date for the heliacal rising of ^{mul}LÚ.HUN.GÁ. The identification of Nisan with Anu is another example of Anu's importance in late-period Uruk and may follow Assyrian tradition where the month is the month of Anu and Enlil.[38] As written, this reference to Nisan as the month of Anu at the end of the line seems to interrupt the natural flow of lines 13–14.

D i 15–16: SAG.KI = *pūtu*, "forehead, brow," which can also be used as a preposition. We take the term to be the body part here, consisting of three stars. Two stars

[33] This writing is again used in D i 15. For unusual orthographies signaling scribal word-play and sign-play, see Wainer (2013).

[34] So the first entry of the barnyard animal list of Urra = *hubullu*, Urra XIII 1 (MSL 8/1 7).

[35] The earliest clear references to Aries as a ram or sheep in Mesopotamia appear in the two *Kalendertext* SpTU III 104 and 105, belonging to the scribe *Iqīša* who lived in Uruk during the late fourth century B.C.E. These two *Kalendertext* associate ingredients to be used in preparing medical remedies with the sign of the zodiac given by the *Kalendertext* scheme. Where the sign of the zodiac is an animal, the ingredients that are listed are taken from the blood, fat, and hair of that animal. Where the zodiacal sign Aries appears in the scheme, the ingredients are sheep blood, fat, and hair, implying that Aries is seen as a ram rather than as "The Hired Man." Unusually, these two texts write the name of Aries using the logogram BAR for Nisan, the first month of the year, which is the month in which the sun is located in the zodiacal sign of Aries according to the schematic 360-day calendar which underlies the *Kalendertext* scheme. This may perhaps suggest that the re-imagining of Aries as the "Ram" was a recent development in Iqīša's time. See further, Steele (2011, 335–8), Steele (2017b), and Reiner (1995, 117).

[36] See, e.g., Neugebauer (1947); Wallenfels (1993, 282–3); Rochberg-Halton (2004b, 67, note to obv. 7); Steele (2006a); Kurtik (2007, 276). Cf. two writings of Akkadian *lumāšu*, "constellation," with the LÚ sign for human beings in "Assurbanipal's Acrostic Hymn to Marduk and Zarpanitu" (SAA 3, 8–9, no. 2, 37, rev. 9).

[37] Steele (2006a).

[38] See Horowitz Alb (57).

in the head of "The Hired Man" are used as Normal Stars: "The Front Star of the Head of The Hired Man" (β Arietis) and "The Rear Star of the Head of The Hired Man" (α Arietis). The three stars in the head of "The Hired Man" in our text are followed by two more stars in its thigh and four stars in its feet in the next line, thus giving us an above-to-below description of the type we might expect for a constellation in human form—in this context, "The Hired Man." Yet, the four stars in the constellation's feet in line 16 may suggest the four-footed classical sheep configuration of classical Aries rather than the traditional Mesopotamian two-footed human.

D i 16: GIŠ.KUN = *rapaštu*, "loin, thigh, haunch, rump." We imagine that this refers to a part of the constellation in sheep form. This term is also used for part of "The Lion" (Leo), another constellation in animal form.[39]

Section B: D i 17–26, ^{mul}UD.KA.DUH.A

This very complex constellation is described in our text as a clothed human figure with two faces: The front face is that of a human being, and the back face is that of a bearded lion. It also has wings, a three-fold element on its ankle and right shin, a right heel, and more.

The reading ^{mul}UD.KA.DUH.A follows modern Assyriological convention and is justified by the Boghazkoi writing ^{mul}*Ka*$_4$*-ad-du-uh-ha* (KUB 4 47 rev. 46). However, TUH might be preferred on the basis of glosses *tu-uh* for DUH in lexical lists: Aa VIII/1 153: $^{\text{tu-uh}}$DUH = *pe-tú-u šá pi-i* (MSL 14 493) and Antagal D 64: $^{\text{tu-uh}}$DUH = MIN (*pe-tu-ú*) *šá pi-i* (MSL 17 204). The constellation's standard Akkadian name remains uncertain. VR 46 no. 1, 41 gives a literal translation of UD.KA.DUH.A, *u*$_4$*-mu na-'-ri*, "the raging storm."

This constellation appears in the list of stars in the three paths in *Mul.Apin*. Parts of the constellation—its Breast, its Knee and its Heel—are given in the list of stars that culminate when other stars rise on certain days in the schematic calendar in *Mul.Apin* I iv 15–20. Four stars from this constellation appear in the standard Late Babylonian list of *ziqpu*-stars: its Shoulder, the Bright Star of its Chest, its Knee, and its Heel. The Shoulder plays an important role in Late Babylonian astronomy as the star that culminates at sunrise on the date of the Spring Equinox.

D i 17: The first half of the line, as far as the *Glossenkeil*, quotes *Mul.Apin* I i 28. The two epithets that follow the *Glossenkeil* are standard epithets of Nergal.[40]

D i 18–19: Here are the first two of many uses of *ṣalam* in D as the technical term for constellations in human form.[41] For this term in ABC and E, see the commentary to Section I A 1.

D i 19–20: The human figure in line 19 is without a beard. The figure with the face of lion in the next line is bearded. Such bearded lions and lion-composites are common in Mesopotamian art. See, for example, Black and Green (1992, 119–22) and that book's frontispiece. The plural adjectives reflect the fact the Akkadian word for face, *panū*, is grammatically plural. The reference to two faces, front and back, suggests a Janus-type figure.

D i 21: The description of the constellation with an open mouth fits its Sumerian name, "The Demon with The Open Mouth."

---

[39] See CAD R 153 and Pingree and Walker (1988, 322) (the rump).

[40] Tallqvist (1938, 393).

[41] The other examples are in D i 23; ii 3, 5, 7–8, 11, 14–15; iii 8.

D i 22: The *kiṣallu*, ankle, fits well here between the *kinṣu* (shin/knee) and *aṣidu* (heel) of the constellation. Why the shin and ankle might be three-fold is uncertain, but one might suspect that this refers to clusters of three small stars.

D i 22–23: The sense of the two signs KI ÚR is obscure. We suppose that the KI.ÚR is a body part; in which case, given the sequence for our constellation thus far (beard, mouth, wings, shin, heel), the KI.ÚR must be lower than the heel and so it must be the foot, a feature of the foot, or an equivalent lower extremity. If this is correct, we suggest that KI.ÚR has the sense of *išdu* (CAD I/J 240 *isdu* 3 f, "lower extremities, stance"), although so far only ÚR = *išdu* is attested with this meaning. If so, then the end of line 22 must introduce a part of the constellation below the right heel, with more information on this body part once available in the now missing first signs of line 23. The end of line 23 would then explain that this last body part is in human form.

D i 24–25: The restoration of these lines rests on *Mul.Apin* I i 28–30:

¶ ^{mul}UD.KA.DUH.A ^{d}U.GUR

¶ MUL *šá ina* ZAG-*šú* GUB-zu ^{mul}ŠAH ^{d}*Da-mu*

¶ MUL *šá ina* GÙB-*šú* GUB-*zu* ^{mul}ANŠE.KUR.RA

"The Demon with The Open Mouth," Nergal.

The star which stands at its right is "The Pig," Damu.

The star which stands at its left is "The Horse."

"The Horse" constellation can be identified with "The Anzu-bird" constellation (see GSL 159, ^{mul} ^{d}IM.DUGUD^{mušen} = ^{mul}ANŠE.KUR.RA; VR 46 no. 1, 20, ^{mul}ANŠE.KUR.RA = ^{d}IM.DUGUD^{mušen}; Kurtik 2007, 59, no. 219–21; and CAD S 334 *sīsû* 2).[42] This identification suggests a transformation of the traditional Mesopotamian Anzu-bird, a lion-headed eagle, into a flying horse (Pegasus) in first millennium Mesopotamian astronomy, and this may represent another example of interaction between Babylonian and Greek astronomy of the type discussed previously for Aries in D i 12–15.

D i 25: The mere suggestion of traces of the tops of signs toward the end of the line may allow for the most tentative restoration *a-n*[*a*] EN.TE.NA (= *kūṣu*) "for cold-weather/winter." This would match comments relating to ^{mul}IM.DUGUD^{mušen} in Assumed EAE 50 and related materials (BPO 2 40 III 11–11d, 74 XVI 10).

Col. ii

Section C: D ii 1–3, ^{mul}MAŠ.TAB.BA.GAL.GAL, "The Great Twins" (restored)

This first portion of col. ii does not begin with DIŠ and a star name and so must continue on from the end of col. i. The reference to a back of a human figure in ii. 3, and the fact that the next entry is for "The Lesser Twins," ensures that these lines must come from the end of the entry for "The Great Twins" (^{mul}MAŠ.TAB.BA.GAL.GAL) that began in the now-missing bottom portion of col. i. For the

[42] Cf. the equation *an-zu-ú* = *si-su-*[*ú*] in the unpublished fragment BM 52938, and the pun in a medical commentary quoted in George (1991, 157), *an-zu-u* : *an-šu-ú*, Anzu = donkey (Sumerian anše).

standard sequence, "The Great Twins" followed by "The Lesser Twins," see also *Mul.Apin* I i 5–6, and VR 46 no. 1, 4–7.

D ii 1. The surviving line opens with the word *hinšu*, a *Glossenkeil*, and then a word written *ku-ut-ta-ú*. The meaning of both words is uncertain, as is also the function of the *Glossenkeil* here. It is possible that the *Glossenkeil* was meant to indicate *hinšu* = *ku-ut-ta-ú*—that is, that the two are synonyms—but it is also possible that our *Glossenkeil* is meant to indicate that the scribe of D is here offering two variants from previous source material: a source(s) that places a *hinšu* in the right hand of "The Great Twins," and a source(s) that read *ku-ut-ta-ú* in this position. If so, then *hinšu* would not explain *ku-ut-ta-ú* and vice versa. What is clear is that both can be carried (*našû*) in one's hands, as is also the case for *hinšu* with "The Great Twins" in ABC Section II, 6 and for *ku-ut-ta-ú* with "The True Shepherd of Anu" later in D ii 16.

The word *ku-ut-ta-ú* itself may be a spelling for what the dictionaries give as *kūtu/kuttu/kutû* (AHw 519, CAD K 611–612), a large can or jug, and figuratively an elephant's trunk. The only other attestation of *hinšu* given in the dictionaries are in two Neo-Babylonian texts, where *hinšu* forms part of the equipment of chariots. In one text, YOS 6 53, gold is used to repair a *hinšu*. CAD H 195 translates *hinšu* as "whip or goad," but the entry in YOS 6 does not seem to fit this description unless what is being repaired is the handle of the whip. AHw 347 and CDA 116 give "ein Werkzeug?" ("an implement?").

Section D: D ii 4–9, ^{mul}MAŠ.TAB.BA.TUR.TUR, "The Lesser Twins"

D ii 4: Again, the translation attempts to convey sign-play. The line begins with the standard Sumerian writing for the name of the constellation with determinative MUL, and it is then followed by two more writings of the name with the divine determinative DINGIR: ^{d}MAŠ.MAŠ.TUR.MEŠ written in Sumerian, and then a writing in which the first element is in syllabic Akkadian ^{d}*tu-ma-mu*^{meš}. D iv 18' repeats much of this line.

Section E: D ii 10–17, ^{mul}SIPA.ZI.AN.NA, "The True Shepherd of Anu"

"The True Shepherd of Anu" appears in the list of stars in the three paths and the list of constellations in the Path of the Moon in *Mul.Apin*. In the so-called TE tablet (BM 77824), "The True Shepherd of Anu" appears alongside "The Great Twins" in the entry for Month III,[43] and the name is occasionally used in Late Babylonian texts as an alternative to "The Twins" for the zodiacal sign Gemini. The "True Shepherd of Anu" did not give rise to any Normal Stars itself, but it appears as a reference point in the name of one Normal Star: ""The Twins' Star Near the Shepherd" (γ Geminorus).

D ii 10: This line parallels *Mul.Apin* I ii 2: ¶ ^{mul}SIPA.ZI.AN.NA ^{d}*Pap-sukkal* SUKKAL ^{d}*A-nim u* ^{d}INNIN, "The True Shepherd of Anu," Papsukkal, the vizier of Anu and Ištar. However, note that Antu replaces Ištar both in our line and in D iv 20 (Papsukkal, the vizier of Anu and Antu), confirming a late synchretism between Ištar and Antu of Uruk. For this, see also the identification of Antu with ^{d}*Nin-si$_4$-an-na*, a name for Venus, in MLC 1890 8 (Beaulieu 1995, with discussion on pp. 201–4). The equation ^{mul}SIPA.ZI.AN.NA = Papsukkal and/or his Sumerian equivalent Ninšubur is repeated numerous times elsewhere, including the Astrolabe menology and star-catalogue.[44]

[43] Weidner (1915).

[44] Alb B I i 38–40, 45–47, Alb B II i 9–11, with further examples in the discussion on Horowitz Alb (67–8). See also the commentary to D iv 11–12 (Section O).

D ii 12–13: The reference to the "The Twins" in front of "The True Shepherd of Anu" quotes *MulApin* I ii 3–4. This same set of twins is also noted among seven twins constellations in GSL 231–236. In the parallel list KAR 142 i 26–34 (Pongratz-Leisten 1994, 221), this set of twins appears as ^{mul}MAŠ.TAB.BA *ša ina* IGI ^{d}PAP.SUKKAL, "The Twins which are in front of Papsukkal": astronomical Papsukkal = the constellation "The True Shepherd of Anu" as in D ii 10. *MulApin* I ii 3–4 is also quoted in a commentary to the diagnostic omen series *Sakikku*.[45]

The identification of this pair of stars with Lulal and Latarak suggests that this pair is the same astronomical twin as ^{mul}LÚ.LÀL and ^{d}*La-ta-rak* = Sin and Nergal in 5R 46, no. 1, 22. For "The Great Twins" identified with Sin and Nergal, see 5R 46 no. 1, 4–5, and perhaps also BM 82923, 14.[46]

MUD = *uppu*, part of the locking device on doors and gates, a very appropriate item to be paired with the *namzaqu*, "the key."

D ii 14: KÁ-*a-ni* may be a writing for the plural of gates, *bābāni*, or the homonymous adverb with the meaning "outside." The former may be preferred on the basis of passages where figurines of Lulal and Latarak are buried at gates to keep out evil, as in the following from *Bit Meseri*:[47]

níg-hul nu-te-gá ^{d}lú-làl ^{d}la-ta-rak ká-ta gub-ba-zu

ana mimma lem-ni ṭa-ra-di ıā ṭehê ^{d}MIN *u* ^{d}MIN *ina ba-a-bi ul-ziz*

(Meier 1941–1944, 150: 211–2)

In order to drive out any evil, (so that) it will not come near, I stationed Lulal and Latarak at the gate

For the second possibility, one might point to the following passage in *The Nippur Compendium*: The Asakku, Latarak, the son of Anu, conquered by Ninurta, whose dwellings are outside (*a-hat*) the city (George 1992, 157, no. 48).

D ii 14–15: Our lines state that only the forward twin is bearded, whereas the second with the face of Latarak is not. It is not clear if this is a scribal omission or if it has to do with the perceived image of Latarak. Another human figure without a beard forms part of ^{mul}UD.KA.DUH.HA in D i 19.

D ii 16–17: The entry for "The Rooster" here is a near parallel to *MulApin* I ii 5, ¶ MUL šá EGIR-*šú* GUB-*zu* ^{mul}DAR.LUGAL, "the star which stands after it" ("The True Shepherd of Anu") is "The Rooster." Note, however, the preposition "below" in our text as opposed to "after" in *MulApin*. For *ku-ut-ta-ú*, see the commentary to D ii 1.

Section F: D ii 18–20, ^{mul}ALLA, "The Crab"

The description of "The Crab" here nearly matches that of ABC Section IV. In both, "The Crab" consists of eleven stars. Here, four stars forming an *apsamakku* represent the two claws of "The Crab" and the constellation's feet, six stars are placed inside the *apsamakku* for the body of "The Crab," and one more star for its head. This configuration matches the drawing of "The Crab" on Section 2 of the Sippar *ziqpu*-star planisphere, where one sees for certain ten stars: single stars at the

[45] George (1991, 150, no. 30b).

[46] Horowitz Alb (147).

[47] See RlA 7 164 in the discussion of Lulal and Latarak by W. G. Lambert.

constellation's four extremities (top and bottom, and right and left), and six stars in three pairs of two inside them—with an eleventh star perhaps just visible for the head on the photograph (Horowitz and Al-Rawi 2001, 176–7). However, the late-Uruk *ziqpu*-star catalogue VAT 16436 11 gives the number of stars in constellations as ten rather than eleven: ^{mùl}KÚŠU 10, "The Crab," ten stars.[48]

D ii 18: The opening words quote *Mul.Apin* I i 7. The wagon box of Ursa Major is also described as an *apsamakku* in ABC Section VII and D iii 14. Á.MEŠ-*šú* at the end of the line may render the same word as ABC Section IV, A 13//C 12' *i-tu-ti-šá*.

D ii 19: The configuration of the six stars belonging to the body of "The Crab" on the Sippar *ziqpu*-star planisphere, arranged as three pairs of two, must be what our sources mean by *ahinnû ahameš rakbū*, "six stars side by side straddle one another." Cf. the syllabic writing of the verb *rak-bu* in the parallel entry in ABC Section IV.

D ii 20: We restore Jupiter standing in front of "The Crab" at the end of the line on the basis ABC Section IV, A 15: [^{mul}S]AG.ME.GAR *ina* IGI-*šú e-ṣir*, "[J]upiter is drawn in front of it." See the previous discussion on p. 29.

Section G: D ii 21–22, ^{mul}KAK.PAN = *Šukūdu*, "The Arrow" (Sirius)

In *Mul.Apin*, "The Arrow" (Sirius) appears only in the list of stars in the three paths. In Late Babylonian astronomy, however, "The Arrow" takes on a much more important role, no doubt in large part because Sirius is the brightest star in the sky. Various mathematical schemes for calculating the dates of the first appearance, acronychal rising, and last appearance of Sirius alongside the dates of the solstices and equinoxes are preserved,[49] and several theoretical texts seem to use the synodic cycle of Sirius as a surrogate for the length of the solar year.[50]

D ii 21–22: ^{mul}KAK.PAN is one of a number of writings for "The Arrow" which is attested alongside the more common ^{mul}KAK.SI.SÁ.[51] The entry for "The Arrow" in Section H is immediately followed by that for "The Bow" in Section I, forming the pair "Arrow and Bow" as in *Mul.Apin* I ii 6–7. The entries in Sections H–I do not parallel those in *Mul.Apin*.

The identification of "The Arrow" with Adad is surprising because there is a very strong tradition identifying "The Bow" with Ninurta as in, for example, *Mul.Apin* I ii 6 and the menology and star-catalogue of Alb B (Alb I ii 1, 8; II i 13).[52] In fact, to our knowledge, "The Arrow" is never identified with Adad elsewhere. Thus, we surmise that the identification of "The Bow" = Adad is meant to facilitate an indirect association between the constellation and Adad's parents, who in our context can be none other than Anu and Antu of Uruk. A similar innovation is to be seen in the identification of "The Bow" with Adad's wife Šala in Section I, thus making "The Arrow" and "The Bow" husband and wife.[53]

The restoration at the end of line 21 yields the common epithet of Adad, *gugal šamê u erṣetim* (see, e.g., Tallqvist 1938, 73–4).

48 Hunger and Pingree (1999, 88); Schaumberger (1952, 226); Fincke and Horowitz (forthcoming).
49 Sachs (1952); Britton (2002).
50 Britton (2002).
51 CAD $Š_I$ 448–451; Kurtik (2007, 243).
52 Horowitz Alb (70–71), with further examples.
53 For Anu in late-Uruk tradition, see Beaulieu (1992).

Section H: D ii 23–24, ^{mul}PAN, "The Bow"

D ii 23: As noted previously, the identification of "The Bow" with Šala, Adad's wife, makes "The Arrow" and "The Bow" husband and wife in our text. Although this is the only example of this married couple in astronomy that is known to us, it does justify the identification of Šala as princess, since her spouse Adad, as "The Arrow" in Section G, is the son of Anu and Antu, the King and Queen of the Uruk pantheon, and so is a prince. Again, one suspects that such innovative identifications are intended to facilitate associations between the stars in the sky and Anu of Uruk and his family members. "The Bow" is normally identified with Ištar as in *Mul. Apin* I ii 7, *Ee* VI 86–93, and Alb B I i 14–16 and parallels.[54]

D ii 24: For this line, cf. the Kassite period precursor to the Alb star-catalogue HS 1897 7: ¶ MUL KI.MIN (*ša* EGIR-*šu* GUB-*zu*) ^{mul}PAN *Iš*$_8$-*tár e-la-ma-tum a-na ṣe-er-re-et ša-me-e*, "the star (which stands after it) is 'The Bow,'" Ištar – the Elamite, for the lead-rope(s)/breasts of heaven.[55] The construction *ana ṣerret* [*šamê*] is of the same type as a number of short astrological comments with *ana* in HS 1897, its late parallel BM 55502, and texts assigned to Assumed EAE 50.[56]

Section I: D iii 1–3, ^{mul}MUŠ = *nirahu*,[57] "The Snake" (Hydra)

Lines 2 and 3 aptly describe elements of this constellation upon which a raven, said to be a star of Adad, stands. This raven is without doubt ^{mul}UGA, "The Raven," which is identified as a star of Adad in *Mul.Apin* I ii 9: ^{mul}UGA *a-ri-bu* MUL ^{d}IM, "'The Raven' is a raven, a star of Adad." This ensures that the constellation named in line 1 must be "The Snake," because "The Raven" is drawn standing on top of, and apparently pecking, this constellation's tail in the micro-zodiac text VAT 7847 + AO 6448 (see plate 1). The traces at the beginning of the star-name in line 1 conform to the reading ^{mul}M[UŠ], "The Snake," but what follows is too damaged to confirm a match with the entry in *Mul.Apin* I ii 8 for this constellation[58] or to suggest something else that fits a snake or any mythological reptile. Yet, the drawing of the snake constellation noted previously does have wings and, as preserved, front feet, matching the description in our line 2. No back feet are currently to be seen, but there is room for the missing rear feet in the current break between the obverse and reverse of the ancient drawing where the snake's body rises to make room for what must have been this missing element. The area between the feet is probably what is called MURUB$_4$ ^{mul}MUŠ, "the middle of 'The Snake,'" in *The GU Text*.

This same sequence, a raven at the end of ^{mul}MUŠ, is also implied in the *Mul. Apin* star-catalog, where "The Snake" occurs in *Mul.Apin* I ii 8 immediately before "The Raven" in *Mul.Apin* I ii 9.

In the drawing on VAT 7847 + AO 6448, to the right of "The Raven" is an eight-pointed star labeled Mercury and the drawing of a female figure holding a barley-stalk (*šubultu*). This same sequence is described in String K of *The GU Text*:

> ^{mul}UDU.IDIM GU$_4$.UTU *šá* ŠÈ ^{mu}[^{l}]AB.SÍN *ina* IGI ^{mul}UGA GUB-*zu* GU
>
> (BM 78181 17–18 [Pingree and Walker 1988, 314–5])

---

[54] Kurtik (2007, 88); Horowitz Alb (74).

[55] Horowitz Alb (103, Ea 6).

[56] See the commentary to D i 10–12.

[57] Partially restored. For the Akkadian name, see Urra XXII 269: ^{mul}muš = *ni-ra-hu*.

[58] ¶ $^{mul\,d}$MUŠ ^{d}*Nin-giš-zi-da* EN *er-ṣe-tu*$_4$.

The planet Mercury which stands toward "Th[e] Furrow" in front of

"The Raven"– a string.

The position of Mercury in this sequence refers to the planet's *bīt niṣirti*. "The Furrow" (^{mul}AB.SÍN[59]) further along the string can be identified with its classical equivalent, the goddess holding the barley-stalk = the constellation Virgo. Here again we find a link between an older Mesopotamian view of the constellation, that of a furrow, with the later classical equivalent—the Virgin with her barley-stalk. This equivalence, however, is probably Mesopotamian in origin, because *Mul.Apin* I ii 10 already identifies ^{mul}AB.SÍN with a goddess and the barley-stalk:[60]

^{mul}AB.SÍN ^{d}*Ša-la šu-bu-ul-tum*

"The Furrow," the goddess Šala, the barley-stalk

In any case, the two registers together yield a sequence of six stellar figures: Jupiter, "The Snake," "The Lion," "The Raven," Mercury, and "The Furrow."[61]

Section J: D iii 4–6, ^{mul}*Ninmah*, Ninmah

D iii 4: The identification of Ninmah as "The Lady of the Gods" is not to be found in the entry for this star in *Mul.Apin* I ii 21, but it is repeated in Urra XXII 243' [^{mul} ^{d}ni]n.mah = $^{\ulcorner d \urcorner}$*be-let* DINGIR.MEŠ, and the Astrolabe text Sm. 1492 4', ^{mul}*Ninmah be-let* DINGIR.MEŠ [. . . (Horowitz Alb 152).[62] The significance of the last five signs in the line continues to elude us. The position of the signs, between the constellation's name and epithet, and the statement that it is a clothed figure suggest that these signs somehow describe the constellation or give an alternative name. Our tentative reading and translation take *pušku* as a container of some sort and A.ME for A.MEŠ (*mû*), "water," with the two forming a construct state. This is better than taking the first as a verbal adjective and the second as a participle, because both forms would be masculine, whereas the goddess Ninmah is clearly feminine as is indicated in the grammar of our text by the feminine form of the statives (*labšat* and *šaknat*) in the very next line.[63] However, we admit that the our tentative first noun seems to have nothing to do with the two attested words *pušku* in Akkadian (CAD P 542–543).[64]

D iii 6: The first word here too is obscure. As written, it must be for something like *dirri*/*tirri*/*ṭirri*/*derri*/*terri*/*ṭerri*, but again no such word is appropriate

[59] Urra XXII 267 now proves the equation ^{mul}ab.sín = *ši-ir-'u*, "The Furrow," rendering earlier suggestions to read the name *abšinnu* and *sissinnu* obsolete.

[60] For "The Furrow" = classical Virgo with the barley-stalk and related matters, see also Hallo (2008, 238–9), Wallenfels (1993, 285) with fig. 8 = Kurtik (2007), fig. 31.

[61] See the commentary to D ii 20 for the possibility of adding "The Crab" to this sequence before Jupiter. For a similar depiction of the corresponding constellations from ca. 1575 Italy—Hydra with Corvus on her tail, Leo, and Virgo—see Hallo (2008, 249 with figures).

[62] Cf. the Astrolabe *mukallimtu* BM 82923 19: ^{mul}*Ninmah* = *šarrat ilāni*, "Queen of the Gods" (Horowitz Alb, 141).

[63] For this same reason, A.ME cannot be read *a-šib*, "he sits, he is sitting, seated."

[64] Alternatives including *pušqu*, *pusku*, *pušqu*, *busku*, and *bušqu* also yield no satisfactory meaning, but one might contemplate a relationship to the adjective *puššuqu*, which occurs once in an Old Babylonian text in the context of water (CAD P 545).

to our context.[65] One may suspect a scribal error and amend the text to *dir-rat*! for the lash of a whip as in ABC Section III and D ii 8 with mulMAŠ.TAB.BA.TUR.TUR and mulE$_4$.RU$_6$ in ABC Section VI.

Section K: D iii 7–12, mulEN.TE.NA.BAR.HUM = *Habaṣīrānu*, "The Mousy One"

The Sumerian name may be best read as mulen.te.na.bar.guz and understood as "The Shaggy Winter Star."[66] For *Habaṣīrānu* as the standard Akkadian equivalent, see, for example, Urra XXII 263': [mu]len.te.na.bar.hum = *Ha-ba-ṣi-ra-nu*.

This section names three sets of stars, all identified with modern Centaurus: mulEN.TE.NA.BAR.HUM itself, the divine pair Šullat and Haniš, and mulNU.MUŠ.DA.[67]

D iii 7: The entry quotes *Mul.Apin* I ii 22. For other identifications of mulEN.TE.NA.BAR.HUM with Ningirsu, see, for example, Alb B II i rev. 8, K. 4292 11 (BPO 2 40 III 5b), and cf. GSL 157: mul*Ha-ba-ṣi-ra-nu* = d*Nin-gír-su*.

D iii 8: The heads of constellations in human form are typically represented by a single star in our group. See, for example, D ii 20, and note the two stars representing the two heads of "The Lesser Twins" in D ii 6–7. Cf. ABC Sections I–IV, VI, and X.

D iii 10–11: These lines can be restored on the basis of *Mul.Apin* I ii 25: ¶ 2 MUL.MEŠ *šá* EGIR-*šu* GUBme-*zu* d*Šullat u* d*Haniš* dUTU *u* dIM, "the two stars which stand after it are Šullat and Haniš, Šamaš and Adad."

D iii 12: This line is to be restored on the basis of *Mul.Apin* I ii 26–27, where the entry for mul*Numušda* there also follows that for Šullat and Haniš:

> ¶ MUL *šá* EGIR-*šú-nu* GUB-*zu* GIM d*É-a* KUR-*ha*
> GIM d*É-a* ŠÚ.MEŠ mul*Nu-muš-da* dIM
>
> The star which stands after them, which like Ea rises,
> like Ea sets, Numušda, Adad.

For mul*Numušda* identified with Adad elsewhere, see Urgud (MSL 11 41, 48) and *ACh Adad* 17, 7:[68]

> [¶ mul]NU.MUŠ.DA d*Marduk* mulNU.MUŠ.DA d*Adad šá ina* ituBÁRA d*Adad* GÙ-*šú né-eh*[69]
>
> Numušda is Marduk, Numušda is Adad, regarding Adad whose thunder abates in Nisan

Section L: D iii 13–23, mulMAR.GÍD.DA, "The Wagon"

D iii 13: This line opens a section relating to mulMAR.GÍD.DA, which continues down to the end of what survives of col. iii. The description of mulMAR.GÍD.DA

[65] One remote possibility may be a word that CAD Ṭ 105 takes as *ṭêru*. This word is attested only once, in Proto-Aa, as an equivalent of Sumerian KAK (read *ga-ag*), and so it may perhaps be for a weapon of some type.

[66] en.te.na = *kuṣṣu*, "cold, frost, winter," bar.guz = *apparrû*, "having wiry(?) hair." For the reading guz note the gloss gu.[uz] in Urra XIII 8 (CAD A II 179).

[67] Kurtik (2007, 133, 202, 393, 508).

[68] For discussion and further examples, see Horowitz Alb (150).

[69] GÙ, for the "roar of Adad" = thunder (*rigmu, šigmu*).

at the beginning of the section parallels that in ABC Section VII. We restore the name of the goddess Ninlil in the missing space at the end of the line in accordance with *Mul.Apin* I i 15 (^{mul}MAR.GÍG.DA ^{d}*Nin-líl*) and Alb B II iii 8–10 (^{mul}MAR.GÍD. DA ^{d}*Nin-[líl]*). For the wagon constellations identified with goddesses including Ninlil, see Horowitz Alb (116). Entries for ^{mul}KA$_5$.A, "The Fox" and "The Ewe" (^{mul}U$_8$ = *immertu*) in *Mul.Apin* I i 15–18 likewise follow those for "The Wagon." The broken reference to ^{mul}GÀM in D iii 16, which parallels *Mul.Apin* I i 4, does not follow the *Mul.Apin* sequence.

D iii 14: For a lexical equation, *eriqqu*, "wagon" = *narkabtu*, "chariot," see *Malku* II 199 (Hrůša 2010, 66).

D iii 16: ^{mul}GÀM = *gamlu*, "The Crook," to be identified with Auriga.[70] The intent of this broken line is for now unclear.

D iii 17: The two faint stars here (*unnutūtu*) are presumably the same as the two lower stars (*šaplūtu*) in ABC Section VII. The end of the line is restored on the basis of ABC Section VII, A 6–7.

D iii 18–19 quotes *Mul.Apin* I i 16–17. The pole of "The Wagon" is here called its *zarû*, a synonym of *mašaddu*, the term that is used in ABC Sections VII–VIII for the poles of both Ursa Major and Ursa Minor. See *Malku* II 209 *za-ru-u* = *ma-šad-du* (Hrůša 2010, 66) and GSL 162 for this equation in an astronomical context: ^{mul}*za-ru-ú* = *ma-šad-du*. Two other named parts of the wagon constellations are the *kalakku* (see the commentary to ABC Sections VII–VIII) and the *ṭurru* (see the commentary to D iv 15' and E 11–12).

D iii 18–21: These lines name two stars in the vicinity of "The Wagon"—"The Fox" and "The Ewe"—both of which to date have not been positively identified. The three occur in the same sequence in *Mul.Apin* I i 15–18.

The reference to "The Fox" appears to be a direct quote from *Mul.Apin* I i 16–17 (¶ MUL *šá* KI *za-ri-i šá* ^{mul}MAR.GÍD.DA GUB-*zu* ^{mul}KA$_5$.A ^{d}*Ér-ra gaš-ri* DINGIR^{meš}) so we may restore the star-name at the end of D iii 18 as ^{mul}MAR.GÍD.DA with utmost confidence. Thus, "The Fox" in both our group and *Mul.Apin* is placed *itti* (with) the pole (*zarû*) of "The Wagon." The questions then are which stars comprise "The Fox" and what does the preposition *itti* mean in this context. Modern Assyriology has suggested a tentative identification between "The Fox" and stars in the vicinity of the pole of Ursa Major,[71] but a more precise identification depends in part on what is meant by *itti* (with). Does this mean that poles of "The Wagon" and "The Fox" share some stars, or that they are totally separate but adjacent to one another, or something else? Unfortunately, philology cannot help us here, because the preposition *itti* is not used elsewhere in our group to describe the relative position of two stars or sets of stars, and no fox-shaped set of stars seems obvious in this portion of the sky.

"The Ewe" in line 21 is also listed on its own in *Mul.Apin* I i 18. In fact, the entry for "The Ewe" in our lines 20–21 begins by apparently quoting *Mul.Apin* I i 18,[72] but the sign after ^{mul}MAR.GÍD.DA, which we read as DIŠ just before D iii 20 breaks off, can hardly be the beginning of GUB. Our text D also differs from *Mul.Apin*

[70] See, e.g., Hunger and Pingree (1999, 272) and Kurtik (2007, 145).

[71] Hunger and Pingree (1999, 272); Hunger and Pingree (1989, 137); BPO 2 12. See also Kurtik (2007, 239).

[72] ¶ MUL *šá ina* SAG.KI ^{mul}MAR.GÍD.DA GUB-*zu* ^{mul}U$_8$ ^{d}*A-a*.

in that it gives the Akkadian syllabic rendering ^{mul}*immertu* for the name of "The Ewe," what *Mul.Apin* gives in Sumerian as ^{mul}U$_{8}$.[73] The writing of the name of the goddess that follows is damaged but appears as ^{d}⌜*e*$^{?}$⌝-*a*; however, this must be for the goddess Aya as in *Mul.Apin* I i 18: . . . ^{mul}U$_{8}$ ^{d}*A-a*.

An identification of the stars comprising "The Ewe" seems dependent on finding a suitable identification for "The Fox."

D iii 22: This next star is "Great Antu of Heaven," which is placed in "The Wagon" in a late ritual instruction text from the Bit Reš in Uruk alongside Great Anu of the Heavens:

> . . . *ki-ma šá* ^{d}60 GAL-*ú šá* AN-*e it-tap-ha An-tu*$_{4}$ GAL-*tu*$_{4}$ *šá* AN-*e ina* ^{mul}MAR.GÍD.DA *it-tap-ha a-na tam-šil zi-i-mu bu-un-né-e* MUL *šá-ma-mi* ^{d}*A-nim* LUGAL *it-ta-ṣa-a ṣa-lam ba-nu-ú* . . .
>
> (Linssen 2004, 245, no. 15–17)
>
> . . . as soon as Great Anu of Heaven appears (and) Great Antu of Heaven in "The Wagon" appears (he will recite) "Likeness of the glow of the most beautiful stars of the sky" (and) "Anu, the King, has come out, the beautiful image" . . .

Great Antu of Heaven cannot be positively identified, but Great Anu of Heaven is identified with ^{mul}MU.BU.KÉŠ.DA, the star that follows "The Ewe," in *Mul.Apin* I i 19:[74]

> ^{mul}MU.BU.KÉŠ.DA ^{d}*A-num* GAL-*ú šá* AN-*e*
>
> MU.BU.KÉŠ.DA, Great Anu of Heaven

Diri Nippur 9 18 (MSL XV 32) offers evidence for Sumerian MU.BU, as in the star-name, being a term for a yoke, Akkadian *nīru*: [x.d]u.ul = [MU].BU = *ni-ru-um*.[75] As such, the Sumerian star-name ^{mul}MU.BU.kéš.da can be understood as "The Hitched Yoke,"[76] and so it is part of the mechanism that connected "The Wagon" to an imagined stellar draft animal. If one follows the direction of the pole of "The Wagon" in the sky, one soon reaches the upper portion of Boötes, and the stars Segginus and Nekkar (β and γ Boötes), which together would be a fitting pair for Great Anu and Great Antu of the Heavens; if connected, they would form a horizontal crossbeam in the sky to which the pole of "The Wagon" could have been thought to have been tied.[77]

Section M: D iv 1'–2'

Almost nothing survives from Section M, the end of which is marked by the horizontal line following col. iv 2'. Given that line 1' does not begin with DIŠ, the

---

[73] Previously, Hunger and Pingree (1989) had rendered the Sumerian ^{mul}U$_{8}$ as *lahru*. See the dictionaries for u$_{8}$ = both *lahru* and *immertu*. The new edition of *Malku* = *šarru* (Hrůša 2010, 140) no longer gives the equation *immertu* = *lahru* for Malku V 34 that is recorded in CAD.

[74] For this same identification, see also 5R 46, no. 1, 12. One must allow for the possibility that *ṣalam* in the passage from Linssen (2004) previously has the same special meaning as "constellation in human form" that we find in our group. If so, we should perhaps be looking for a human figure in ^{mul}MU.BU.KÉŠ.DA.

[75] Perhaps to be restored as [šu.d]u.ul, giving MU.BU a reading šudul$_{x}$, and thus a homonym for standard šudul = *nīru*.

[76] kešda = *rakāsu*, "to tie, to attach something to something else."

[77] This would also fit the identification of "The Harness" constellations (^{mul}*Nattullu*) with stars in Boötes (Kurtik 2007, 361; Hunger and Pingree 1999, 273). For parts of wagon-constellations and stars in Boötes, cf. the commentary to D iv 15 (see pp. 59–60).

section must have begun higher up on the original tablet, but how much is missing of Section M cannot now be determined. If, as we believe, Section N concerns "The Great One" (= Aquarius), Section M should concern a nearby constellation in the region of Capricorn.

Section N: D iv 3'–6'

This badly preserved section can be tentatively identified as describing "The Great One" (^{mul}GU.LA, Aquarius) on the basis of the feature *qup-pe-e* in D iv 6'. The constellation "The Great One" appears occasionally in observational texts and lends its name to the zodiacal sign Aquarius in the late period. Two entries in the list of Normal Stars on BM 36609+ refer to "The Great One," both of which mention a feature named *qup-pu* (one at the front of the Great One [*qup-pu* IGI *šá* ^{múl}*Gu-la*],[78] the other at the rear [*qup-pu* ÁR]). These two stars also appear occasionally in Astronomical Diaries and other observational texts.[79] No other constellations are known with this feature. The identification of this section as referring to "The Great One" is supported by the realization that this section is the last part of this text to provide descriptions of the constellations; "The Great One" would complete a circuit in the sky that started with "The Field" in Section A.

D iv 6': The writing *qup-pe-e*, with the final e-vowel indicated as long, would seem to demand that the word be taken from CAD Q 311–312 *quppû*, "a knife," rather than CAD Q 307–310 *quppu* A, "a wicker basket or wooden chest, box." In BM 36609+, the object in question is written *qup-pu* and in the observational texts it is written either *qup-pu* (Diary No. −123 Rev. 15, Planetary Text No. 60 Obv. I 5') or *ku-up-pu* (Diary No. −366B 6), both of which could be taken as having either a short or a long final u.

Section O: D iv 7'–14', ^{mul}KU$_6$, "The Fish"

D iv 7': This line begins like the entry for "The Fish" in *Mul.Apin* I ii 19:

> ¶ ^{mul}KU$_6$ ^{d}*É-a a-lik* IGI MUL^{meš} *šu-ut* ^{d}*É-a*
>
> "The Fish," Ea, the forerunner of the stars of Ea

However, the trace after Ea's name does not fit *Mul.Apin*'s *a-lik* .... For this standard identification of "The Fish" constellation with the god of the waters Ea, see Horowitz Alb (98).

D iv 8'–10': Barely visible traces at the end of D iv line 8' suggest that Saturn's name ^{mul}GENNA (TUR.DIŠ) is used here. A writing ^{mul}SAG.UŠ would have to be very squeezed to fit the space, and ^{mul}*zi-ba-ni-tum* is much too long. The appearance of ^{mul}UDU.IDIM before the entry for Mars is peculiar. UDU.IDIM is sometimes added to the beginning of the names of Mercury and Saturn (e.g., ^{mul}UDU.IDIM.GU$_4$.UD and ^{mul}UDU.IDIM.SAG.UŠ in *Mul.Apin* II I 5–8), but it is not normally found as part of the name of Mars. Furthermore, if UDU.IDIM was part of the name of Mars, we would not expect to find a second MUL determinative.

This most likely restoration of the names of the seven planets is based on what appears to be a standard sequence for Late Babylonia—Jupiter, Venus, Mercury,

[78] The photograph confirms the reading MÚL, not DINGIR as read by Roughton et al. (2004, 568).
[79] See Jones (2004, 488) for a list of occurrences.

Saturn, and Mars plus the Sun and Moon—that is used in texts such as the Goal-Year Texts, the Astronomical Diaries, and in many astrological works. Cf. also the list that is embedded in the ritual instruction text for the daily rituals of the Reš Temple at Uruk:

> . . . ^{d}60 *ù An-tu*$_4$ *šá* AN-*e* ^{d}SAG.ME.GAR ^{d}DILI.BAD ^{d}GU$_4$.UTU ^{d}GENNA ^{d}*ṣal-bat-a-nu* KUR-*ha* ^{d}UTU *ù* IGI.DU$_8$.ÀM ^{d}30 . . .
>
> (Linssen 2004, 175, no. 30–31)[80]
>
> . . . Anu and Antu of the Heavens, Jupiter, Venus, Mercury, Saturn, Mars, the rising of the Sun and appearance of the Moon . . .

Earlier lists of the seven ancient planets give a different sequence: *Mul.Apin* I i 1–6, Sun, Moon, Jupiter, Venus, Mars, Mercury, Saturn; GSL 242–244 and *Antagal* G 303–309 (MSL 17 229), Moon, Sun, Jupiter, Venus, Saturn, Mercury, Mars.[81]

How the list of planets might be connected to the entry for "The Fish" that starts this section is unclear, as is also the reasoning behind the use of DIŠ signs to mark the start of both D iv 8'–9'. This is anomalous as DIŠ is used elsewhere in our group only to mark the beginnings of sections following horizontal rulings.[82] In contrast, lines 11'–14', down to the end of the section, are not marked by DIŠ. Here the author identifies the planets as sons of Anu, followed by further discussion related to Anu.

D iv 11'–12': The surviving parts of the last two signs in line 11' can hardly be for anything else but the name of Papsukkal, who is vizier of Anu and Antu at late-period Uruk.[83] This is also made clear in our text in D ii 10 and D iv 19'–20' where ^{mul}SIPA.ZI.AN.NA is Papsukkal, the vizier of Anu and Antu. It is not clear why Papsukkal is introduced in D iv 11'–14', and how he may be connected to the Igigi. The Igigi, which typically number seven, are here almost certainly to be identified with the just mentioned seven planets although this association is not made explicit in our text, and not known to us from any other cuneiform source. If so, the last part of Section O may be intended to make an association between Anu's abode in the sky above,[84] and his abode on earth, the Bit Reš in Uruk.

D iv 13': The context demands that *rab-biš* be a writing for *râbiš*, "magnificently, abundantly," rather than for the near homonym *rabbiš*, "gently, softly."

Section P: D iv 15'–20'

Section P does not describe the drawing of constellations, but instead seems to show a general interest in Anu, and stars that may be identified with his family and court. This also seems to be the case for the latter portion of Section O where we find mention of the seven sons of Anu and his vizier Papsukkal. The multiple associations between viziers, the "twins" constellations, "The True Shepherd of

---

[80] Note in the next section of the ritual: ^{d}60 *ù An-tu*$_4$ *šá* AN-*e ù* ^{d}UDU.IDIM^{meš} *7-šú-nu*, "Anu and Antu of the Heavens, and the planets, the seven of them" (Linssen 2004, 175, no. 33).

[81] Cf. SAA 2 29 13–15 (*The Succession Treaty of Esarhaddon*): Jupiter, Venus, Saturn, Mercury, Mars, "The Arrow" (Sirius). See Jones and Steele (2011) for a recent overview of planetary orders in Babylonia.

[82] See Watson and Horowitz (2011, 135–7, 150–1) for the use of DIŠ and horizontal rulings to mark entries and sections of text in *Mul.Apin*.

[83] Beaulieu (1992, 58–63).

[84] For "Abode of Anu," *šubat Anim* as a name for heaven see Horowitz (2011, 228).

Anu," and finally Papsukkal, the vizier of Anu and Antu, at the end of the section must ultimately emanate from associations between "The True Shepherd of Anu" and Anu's viziers (see the commentary to D ii 12–13).

D iv 15': For this line, cf. *MulApin* I i 21–22 which names ^{mul}IBILA.É.MAH as the star in the *ṭurru* of ^{mul}MAR.GÍD.DA.AN.NA, "The Wagon of Heaven" (the Little Dipper), and identifies the star as a son of Anu:[85]

MUL *šá ina ṭur-ri-šú* GUB-*zu* ^{mul}IBILA.É.MAH DUMU *reš-tu-ú* ^{d}*A-nu-um*

The star which stands in its ("The Wagon of Heaven"'s) *ṭurru* is "The Heir of The Sublime Temple," the first-ranked son of Anu

In our text, the same star is identified with Ea. Thus, if one then puts these traditions together, we find that we have Ea as a son of Anu,[86] a tradition also known from *Enuma Elish*.[87] In a parallel to our line from Neo-Assyrian times "The Heir of The Sublime Temple" is identified with "The Wagon" (not "The Wagon of Heaven"), and is a son of Enlil, not Anu:

ÉN DUMU.UŠ É.MAH DUMU.UŠ É.MAH *aplu rabû ša Enlil attama*

ištu Ekur [*t*]*uridamma ina qabal šamê itti* ^{mul}MAR.GÍD.DA *tazzaz*

BAM 542 iii 13–16 and dupl. (Reiner 1995, 20, no. 71)

Incantation: "Heir of the Sublime Temple," "Heir of the Sublime Temple," great heir of Enlil are you.

From the Ekur [you] come down here, in the midst of the heavens you stand with "The Wagon"

For the *ṭurru* of "The Wagon" (the Big Dipper) see also Alb B II iii 11–12:

¶ MUL *ša i-na ṭu-ri-ša* GUB-[zu] <MUL> SA$_5$ *i-na pu-ut Ni-ru* ^{d}*E*[*n-líl*]

The star which stand[s] in its ("The Wagon"'s) *ṭurru*, the red <star> at the frontside of "The Yoke," E[nlil].

Thus, the *ṭurru* is another of the named parts of wagon constellations, the others being *mašaddu*, "the pole"; *zarû*, "the cart-pole"; and the *kalakku*, the wagon-box.[88] The entry *Malku* VIII 66 (Hrůša 2010, 424) *durgallu*, reed rope = *ṭurru*, suggests an identification with CAD Ṭ 164–165 ṭurru A, "yarn, twine, wire, string, band," in this case perhaps some sort of ropes or straps that connected "The Wagon" and "The Wagon of Heaven" constellations to their imagined draft animals. Thus, CAD Ṭ 164 ṭurru A and the astronomical feature CAD Ṭ 166 ṭurru C may both be the same word.[89]

Where these ropes or straps may have been located on the stellar wagons is uncertain. It seems most unlikely that the *ṭurru* can be counted among the seven stars of "The Wagon" and "The Wagon of Heaven" which belong to the constellation's pole and wagon box. So, perhaps, we should look farther forward, or to the side of the wagon pole. In the case of the *ṭurru* of "The Wagon" (= Ursa Major, the Big

[85] For more on the *ṭurru* of constellations, see further below.

[86] I.e. (D iv 15'), Ea = ^{mul}IBILA.É.MAH, ^{mul}IBILA.É.MAH = a son of Anu (*MulApin*).

[87] See, e.g., *Ee* I 15–16.

[88] See ABC Sections VII–VII, D iii 18–19 and also *MulApin* I i 17–18.

[89] Cf. *The GU Text*, which makes use of GU, "string," as an astronomical term.

Dipper), one might consider stars in Boötes (^{mul}ŠU.PA), which might have been thought to connect the pole of "The Wagon" with Arctarus, particularly because ^{mul}ŠU.PA is sometimes said to be an astronomical yoke and includes among its constellation parts one known as *nattullu*, "the harness."[90] Similarly, in the case of "The Wagon of Heaven" (Ursa Minor, the Little Dipper), one might look for stars at the front of its pole by Polaris, which Hunger and Pingree (1999, 273) suggest itself might very well be our ^{mul}IBILA.É.MAH, "The Heir of the Sublime Temple."

D iv 16'–19': These lines refer to at least three sets of astronomical twins: 1) "The Great Twins," 2) "The Lesser Twins," and 3) a third set said to "stand in front." The first two have been described earlier in our text: D ii 1–3 (Section C) give what remains from the end of a section for ^{mul}MAŠ.TAB.BA.GAL.GAL and D ii 4–9 (Section D) immediately follows with the description of ^{mul}MAŠ.TAB.BA.TUR.TUR. In fact, D iv 18' repeats D ii 4 almost verbatim:

> ii 4: ¶ ^{mul}MAŠ.TAB.BA.TUR.TUR ^{d}MAŠ.MAŠ TUR.MEŠ ^{d}*tu-ma-mu*^{meš} TUR.MEŠ
>
> iv 18': ^{mul}MAŠ.TAB.BA.TUR.TUR ^{d}MAŠ.MAŠ ^{d}*tu-ma-mu*^{meš}

The third set of twins in D iv 18'–19' is most likely the same set as "The Twins" in front of "The True Shepherd of Anu" in D ii 12–13.

D iv 16'–17': The discussion of "The Great Twins" in D iv 16'–17' quotes *Mul.Apin* I i 5 which also identifies "The Great Twins" with Lugalgirra and Meslamtaea. *Mul.Apin* I i 6 then goes on to identify "The Lesser Twins" (^{mul}MAŠ.TAB.BA.TUR.TUR) with Alammuš and Ningubluga, but this identification is not repeated in our text.[91]

The Colophon (col. v 1'–8')

The surviving lower half of Source D col. v gives the last eight lines of the tablet's colophon. If this colophon is of the standard type found on late-Uruk tablets, then we indeed have most of the colophon intact, leaving plenty of room for up to as many as twenty or more lines of text above in the missing top portion of the column. The colophon itself indicates that the uranology text was copied from a previous source. This is also indicated by the fact that the colophon does not actually end our tablet. Instead, one finds below the colophon, still in col. v, vacant space down to the bottom edge, with text resuming at the start of what remains of col. vi. Thus, the text in col. vi must have been copied by our scribe from something other than the uranology work now represented by our D cols. i–iv and its colophon in col. v. This opens some interesting possibilities, which we will turn to in our commentary to col. vi.

D v 3'–4': The scribe's name is lost, but he is the son of *Anu-aḫ-ušabši*, descendant of *Ekur-zākir*. Only one known father fits the timeline for a son active in writing tablets in year 97 of the Seleucid Era, but he had at least four sons (see Ossendrijver 2011, 216).

D v 5': Here note the late writing *šešgallu*/*šešgullu* with GU, which is common in late-period Uruk colophons.

---

[90] Kurtik (2007, 365) and cf. a related discussion of parts of "The Wagon" in the commentary to D iii 22 (see p. 55). See also Horowitz Alb (116, no. 770).

[91] For various permutations of the names of "The Twins" constellations identified with various pairs of deities, see the discussion in Horowitz Alb (114).

D v 5'–6': The title "recorder of Enuma Anu Enlil" is previous unknown. This is best explained by the identification *šassukku* as a synonym of *ṭupšaru*, "scribe" (CAD Š II 145), making the individual hold the same title as the more common "scribe of *Enuma Anu Enlil.*"

D v 6': This sense of *piqittu*, that Adapa is responsible for the *Enuma Anu Enlil* series, expands of some on the shades of meaning already available for the word in CAD P 391 piqittu 4–5. We admit that we cannot offer a direct parallel to justify our translation, but we refer the reader to the *Catalogue of Texts and Authors*, which attributes *Enuma Anu Enlil* directly to Ea (Lambert 1962, 64–5; K. 2248 1) and then makes mention of a text with the broken name [UD AN ^{d}EN.LÍ]L.LÁ that is attributed to Ea's acolyte, Adapa-Oannes (K. 2248 5–7). We agree with those who see also in this second name the canonical *Enuma Anu Enlil*, despite Lambert's original explanation to the contrary, that what is meant here is a text UD.SAKAR.AN.^{d}EN-LIL that is named as such in *The Verse Account of Nabonidus* (Lambert 1962, 70). We disagree with Lambert and instead agree with Machinist and Tadmor (1983, 146–50) who argued that *Enuma Anu Enlil* appears twice in the beginning of the catalogue: first as a divine composition composed by the god Ea; and second as the very same composition, now revealed to the human race by Adapa in Ea's name. *The Babyloniaca of Berossus*, as we now know it, offers perhaps the clearest example of a popular Late Babylonian tradition that Oannes (= Adapa) was the main conduit for transmission of knowledge from God to Man in late tradition.[92] We believe that this knowledge apparently included, in the mind of the scribe of our Text D, Text D itself.

Adapa, of course, was the wise sage and *apkallu par-excellence* in ancient Mesopotamia, and an important figure in the writings of the learned scribes of Late Babylonia, particularly those of Uruk.[93]

D v 7': As written, the two signs following ⸢^{lú}*um*⸣-[*m*]*án* are of identical shape, height, and length, both being perfectly acceptable ME signs of the type we have already seen in our manuscript with the horizontal stroke beginning at or just after the vertical stroke.[94] Thus, although it is tempting to read here BAR ME KI for the name of Borsippa (*Bar-sip*^{ki}), this cannot be accomplished without emending the text, which we are reluctant to do without a ready explanation connecting our text, or Adapa (a hero of Uruk), to that city.[95] Given the above, we feel obligated to attempt to read the opening of the line ⸢^{lú}*um*⸣-[*m*]*án* ME ME KI and to seek an explanation for this reading even if we may be accused of proverbially "looking for zebras instead of horses."

We suggest that what we have here is scribal sign-play allowing for two alternate epithets of Adapa:

1) Adapa as *ummân nēmeqi*, "Sage of Wisdom," realized phonetically with metathesis as *ummanmēmeqi*, apparently with the now-coupled double-consonant *nm* compensating for the loss of length in the preceding A-vowel in *ummân*.

---

92 Burstein (1978, 13–4).

93 For an overview, see Sanders (1999, 91–137). Adapa is also named on Sm. 1492, a fragment in the Astrolabe group; see Horowitz Alb (155–6).

94 D i 11, iii 4, iv 8'. In contrast, cf. the BAR/MAŠ signs in D iv 16'–18' (Section P) in the names of "The Twins" constellations. Here we find typical BAR/MAŠ signs with the head of the horizontal stroke placed well to the left of the vertical, yielding the expected form.

95 One could take Borsippa to be part of the date formula; i.e., the tablet was written in Borsippa on the 23rd of the month *Kislīmu*, Year 97 (of the Seleucid Era). However, there is no reason to believe that our tablet comes from Borsippa instead of Uruk, particularly given the numerous references to Anu in the text and total absence of Marduk and Nabu.

2) An even more convoluted writing for *ummân šamê u erṣetim*, "Sage of Heaven and Earth," with ME ME = *šamû u erṣetum*, as is attested in some learned and/or playful writings elsewhere,[96] and with KI (= *erṣetum*) here as a sort of learned hint or complement.

In fact, the possible presence of the scribal anomalies noted previously may strengthen the case for reading ME ME KI rather than *Bar-sip*^{ki}. It has been observed that such irregular or unexpected writings may have been intended to alert readers of the presence of scribal word- or sign-play.[97]

The date given for the writing of the tablet, "The month of *Kislīmu*, 23rd day, Year 97 (of the Seleucid Era)," corresponds to 4 January 214 B.C.E.

D col. vi

Col. vi concludes our tablet, but it does not belong to our Text D in col. i–v, which ends with the colophon. This is also made clear on the physical tablet by the double horizontal ruling after the colophon, the open space left by the scribe at the bottom of col. v, and the fact that col. vi is separated from col. v by a much wider vertical band than those between the previous columns. At present, what is left of col. vi gives seven partially preserved lines that end with a horizontal ruling followed by blank space down to the end of the column, and so the reverse of the tablet. If we assume that our scribe began the text on col. vi at the top of the tablet, this would mean that rev. vi originally yielded somewhere around 35–40 lines of writing when complete. Unfortunately, the contents of the surviving portion of col. vi cannot be identified, although at least parts of tens of signs survive in this section. The very top surface of most of the column is worn away, rendering almost all the signs, most frustratingly, illegible.[98] Given that col. v marks the end of the text belonging to our uranology group, as noted previously, col. vi must contain something from a different text. Thus, our tablet is a sort of *Sammeltext* or, more appropriate in this case, perhaps a scribal workbook containing two separate works (see chapter 1): The uranology text and something else, but what? This could be anything that caught our scribe's eye that could be copied in the space left available on the tablet. If so, col. vi could be from almost anything in the scribal corpus which late-Uruk period scribes were prone to copy, although one might guess that the topic of col. vi was somehow related to the stars.

[96] Horowitz (2011, 229).

[97] Wainer (2013).

[98] P.A. Beaulieu, E. Frahm, and W. Horowitz have all attempted to read and copy this column with no tangible success.

CHAPTER 4

A Three-Constellation Version

Edition

E. MLC 1884. Photographs: plate 15; copy: plates 16–17.

REVERSE

1'. [x x] ⸢x x⸣ [x x x x x x]
2'. [o] ⸢x (x)⸣ [x (x)]
3'. ⸢*ṣa*⸣*-lam šu-ú lu-bu-uš-tu*$_4$ *la-b*[*i-iš*]
4'. *ziq-nu za-qin-nu ku-ur-ku-ru šá-kin*
5'. DIŠ+*en* MUL *ina ku-ur-ku-ru-šú* GUB-*zu*
6'. *ina* ŠU^{II} 15-*šú* GÀM *na-ši*
7'. *i*[*na*$^?$](erased?) GÀM DIŠ+*en* ^{mul}GÀM GUB-*zu*
8'. SAG.DU GÀM SAG.DU UDU.NÍTA
9'. ⸢2(or 3)⸣ MUL *ina* SAG.DU GÀM *e-ṣir*.MEŠ
10'. [x M]UL *ina rit-ti* GÀM GUB-*zu*
11'. [M]UL SA$_4$ *ina ur-šú* GUB-*zu*
12'. [x]⸢(x) x⸣-*šú*$^?$ ^{d}NE$^?$.GI$^?$ *na-áš-rap na-ši*
13'. [x x x] ⸢x x⸣ *ina*$^?$ *na*$^?$*-áš*$^{??}$*-rap e-ṣir*.MEŠ

14'. [MUL.MUL] ⸢^{d}IMIN.BI⸣ DINGIR.MEŠ GAL.MEŠ
15'. [GU$_4$.AN.N]A ⸢*is le-e*⸣ *a-lu-ú*
16'. [x] ⸢x x (x)⸣[E]N$^?$.TE.NA.BAR.HUM
17'. [x] *ina*$^?$ *šap*$^?$*-li*$^?$ ⸢x x (x) $^{mul?}$⸣*nu-mu*[*š-da*]
18'. [x (x) *a*$^?$]-*ge*$^?$*-e* ⸢AN ŠEŠ.MEŠ$^?$⸣ *ki-lal-la-n*[*u*](or: *a*[*n*])
19'. ⸢x x⸣ *pa-ni* UR.MAḪ *la-an-nu šá ṣa-la*[*m*]
20'. ⸢x x x x x x x x x⸣ [x]

remainder lost

Translation

REVERSE

1'–2'. (traces only)
3'. [A hu]man figure it is, dress[ed],
4'. bearded, set with a *kurkurru*.
5'. 1 star stands in its *kurkurru*.
6'. In its right hand it carries a crook.
7'. In the crook a (star called) "The Crook" stands.

8'. The head of the crook is the head of a ram.
9'. 2(or 3) stars in the head of the crook are drawn.
10'. [x] star(s) in the handle of the crook stand.
11'. A bright [st]ar in its shaft stands.
12'. Its [x], fire, bearing a *naṣrapu.*
13'. [. . .] in the *naṣrapu* are drawn.
14'. ["The Stars,"] the seven gods, the great gods.
15'. ["The Bull of Heave]n," "The Jaw of the Bull," a bull
16'. [. . .] . . . [E]N.TE.NA.BAR.HUM.
17'. [. . .] below [. . .] Numušda.
18'. [. . . c]rown, divine brothers, both
19'. [. . .] the face of a lion, torso of a human fig[ure]
20'. (traces)

Commentary

E Section I: rev. 1'–13'

Much of Section I considers "The Crook," but this is not the primary star whose name is now missing. Given that the "The Old Man" is located immediately in front of "The Crook" in the sky and that it is also one of the few stars to possess a *kurkurru,* it seems certain that the opening lines refer to this constellation. Furthermore, in ABC Section I, "The Old Man" is also said to be a clothed human figure with a beard.

Rev. 4'–5': The *kurkurru* of "The Old Man" is also noted in SAA 8 380: 1, which contains the protasis: DIŠ [mu]ŠU.GI *kur-kur-ru-šú i-nam-bu-uṭ,* "If '[th]e Old Man,' its *kurkurru* shines."

Rev. 7'–11': "The Crook" here consists of three parts, its head in the shape of a ram, its handle composed of two or three stars, and what must be its shaft. The terms used for the handle and shaft when applied to the crook are in the context of humans: the hand and the arm (see CAD U/W 266 ūru C).[1] For the handle of "The Crook," see AO 6478 rev. 4, 6: MUL *rit* GÀM; Hunger and Pingree (1999, 85–6) (XVII) identified as θ Aurigae.

Rev. 11'–13': These two lines refer to a further feature of this group. Due to the break at the beginning of line 11', the intention of the author is difficult to fathom. However, a group of stars named *naṣrapu* appear in *ziqpu*-lists and are positioned next to "The Crook." No description of a *naṣrapu* survives. It is unclear whether NE.GI refers to fire in general or more specifically to the fire god Girra.

E Section II: rev. 14'–20'

This section opens in lines 14'–15' with a parallel to the entry for "The Stars" constellation (Pleiades) in *Mul.Apin* I i 44, followed by a reference to "The Bull of Heaven," the constellation after "The Stars" in *Mul.Apin* I ii 1. Lines 16'–17' presents some difficulties. The surviving text certainly suggests restoring the names of the two constellations EN.TE.NA.BAR.HUM and Numušda, but both are located completely across the sky from "The Stars" and "The Bull of Heaven." Perhaps this is the intent here, and *ina šap-li* in line 17' refers to this in some way.

[1] Can this word be related to CAD U/W 260 urû?

Another possibility is that the two constellations are introduced here because they have features that can be compared to those of the bull, although what these might be escapes us.

We understand lines 18'–19' to refer to the moon. The reference to a pair of "divine brothers" would seem to relate to the moon and the sun in this context. A contemporary example is found in the new year's ritual from Uruk MLC 1873 34 (Linssen 2004, 210). The moon's *bīt niṣirti* is located in Taurus between "The Stars" and "The Bull of Heaven." The surviving text and traces allow for a description of the moon as a crown as is common in Mesopotamia,[2] followed by a reference to a figure with face of lion but body of a human figure. This fits, more or less, what we find on the drawing on VAT 7851 where the face of the moon is placed between "The Stars" (Pleiades) and a drawing of "The Bull" (Taurus). Here, the face of the moon is filled by three elements: a human figure, a weapon that he holds in one hand, and what looks like a composite being with the face of a lion (see plate 1), thus matching the description in line 19'. A number of different, but related, ancient Mesopotamian perceptions of the face of the moon are collected and studied in Beaulieu (1999), and a reference to a dagger and lion in the moon is to be found in the mystical-explanatory work KAR 307 (ibid., 93).[3]

[2] See, e.g., *Ee* v 14, Livingstone (1986, 23, 39–40), Stol (1992), and CAD A_I 156.

[3] For a review of the wide variety of what different cultures see/saw in the face of the moon, see, e.g., Krupp (1991, 54–78). A parallel to the tripartite Babylonian view (human figure, lion, and dagger; i.e., human figure, animal, object), may be found in the Gwich'in tradition of "The Boy in the Moon," where one finds in the face of "The Moon" a boy, his puppy, and something that the boy carries on his back. The astronomical traditions of the Gwich'in are currently under study by a group including G. Holton and C. Canon of The University of Alaska, Fairbanks, and W. Horowitz with A. Andre and I. Kritsch of the Gwich'in Social and Cultural Institute of the Northwest Territories, Canada (see Horowitz forthcoming).

CHAPTER 5

The Cuneiform Uranology Texts: Looking Forward

The publication of our group, including the previously known text of Weidner (Source A) and the four new sources B, C, D, and E brings us ever closer to an elusive end-goal of the study of cuneiform astronomical texts: a full mapping of the Mesopotamian star-heavens. This endeavor is now more than 100 years in the making, going back to the decipherment of cuneiform and the early studies of cuneiform astronomical texts of the late 1800s and early 1900s. Recent advances to this goal include new publications of almost all major cuneiform astronomical works including *Mul.Apin* and the *Astrolabes* (*The Three Stars Each*; Horowitz Alb), *The GU Text* and the *DAL.BA.AN.NA Text*, and the star-list in Urra = *hubullu* XXII, all of which are discussed in the introductory materials to our book. More generally, the Russian language reference work on star-names, Kurtik (2007), now offers an updated encyclopedia-type tool for studying the cuneiform starry sky, at least for those who can read Cyrillic or who can handle the transliterations of cuneiform sources in Latin script to mine Kurtik's work for their own purposes. Thus, for now, we can confidently identify many Mesopotamian star-names with individual fixed-stars, and the classical and modern constellations in general use today. Yet, even a cursory glance at Kurtik's book or, for example, the "Catalogue of Constellations and Star-Names with Tentative Identifications" in Hunger and Pingree (1999, 271–7) demonstrates that identifications of many ancient star-names with modern stars and constellations is not yet possible. If so, how has the publication of *The Cuneiform Uranology Texts* advanced us to our goal, what remains to be done—or perhaps what now seems possible—and what as yet seems to us to be still over the intellectual horizon?

First of all, the cuneiform uranology texts have provided a bridge between the mainstream astronomical cuneiform text traditions of the Neo-Assyrian period—for example, *Mul.Apin*—and the small group of drawings of constellations on the later micro-zodiac texts. As argued in the Introduction, the star-names and pictures on the micro-zodiac texts match the constellations as they are known from *Mul.Apin* and other sources that predate the fall of Assyria. Likewise, the drawing of the moon and planets Mercury and Jupiter in their *bīt niṣirti*, "secret place," provides a link between the drawings and Neo-Assyrian royal inscriptions and contemporary *Enuma Anu Enlil* tablets. Thus, it is clear that the basic outline of the Mesopotamian sky as known in the Persian and Hellenistic period was already in place centuries before regular contact is documented between the cuneiform writing and Greek-speaking worlds. This observation is confirmed by the fact that almost all the stars in our exemplars C, D, and E from Late Babylonian period Uruk are already listed in the *Mul.Apin* star-catalogue, which perhaps has its origins in the late second millennium, and even more graphically because our Source D from late-period Uruk provides some passages which seem to provide information that is meant to coordinate between earlier (Mesopotamian) views of constellations and later ideas that are usually assumed to originate in the classical world. The best example of this phenomenon is the discussion of Aries in D i 12–16, which shows knowledge of the constellation as both a human figure ^{mul}LÚ.HUN.GÁ (the Mesopotamian "Hired-Man") and a sheep (as with classical Aries). Likewise, the earlier discussion of Babylonian Anunītum and "The Swallow" joined at their tails points us in the direction of the two Babylonian constellations joined to form classical Pisces, the classical fish-constellation, as opposed to Mesopotamian ^{mul}KU $_6$, "The Fish" = Piscis Austrinus.

Another important contribution of our uranology group is that it often provides the number of fixed-stars that compose constellations and constellation parts. These can be matched with a small group of cuneiform texts that provide drawings of the constellations with individual stars as dots, including the planisphere CT 33 10 (= K. 8538) and the Sippar Planisphere (Al-Rawi and Horowitz 2001). The number of dots assigned to constellations on the latter matches VAT 16437 and a newly identified duplicate that counts the number of fixed-stars in *ziqpu*-star constellations and constellation parts (Fincke and Horowitz, forthcoming). Such correspondences offer definitive proof for a continuity of tradition over the time boundary between the era of the native Mesopotamian sovereignty of the Neo-Assyrian and Neo-Babylonian periods, and political domination by foreign cultures—the Persians and then the Greeks. Thus, we cannot assume a priori that anything new in the Persian or Hellenistic period in cuneiform astronomy was introduced by foreigners. We may very well need to rethink issues relating to the circulation of Mesopotamian and Greek astronomical traditions.

More generally, the Uranology group of texts confirms long-standing identifications of Mesopotamian star-names with their modern and classical equivalents. "The Wagon" and "Wagon of Heaven," with their four-sided front features and poles in Source A, remain undoubtably our Big and Little Dippers, whereas "The Raven" standing on the tail of "The Snake" in both Source D and one micro-zodiac drawing (see plate 1) must be identified, as has long been known, with modern Corvus and Hydra. On such occasions, our Uranology texts are useful in that they confirm that our long-standing consensus view of the Mesopotamian sky is basically correct.

The devil, of course, is in the details. Or more precisely, the devil is in questions such as "Which exact stars form which ancient constellations and constellation parts?" or, from another perspective, in identifying all known cuneiform star-names with their counterparts in the sky—and pinning down the precise meaning of all terminology. More general problems also persist. For example, how do we reconcile the system of drawing stars that make up constellations as points with the free-hand drawings of the constellations as found on the micro-zodiac tablets?

Solving such problems, and other remaining issues pertaining to mapping the ancient Mesopotamian star-heavens, will require further work. One methodology that comes to mind is to start by generating a star-map for ancient Mesopotamia for a given year, let's say 500 B.C.E. We begin by filling in the names of stars, constellations, and constellation parts of which we are certain, and then use our existing resources to try and complete this picture. A similar methodology is implied in

BPO 2 6, in which E. Reiner and D. Pingree write of testing their identification of Mesopotamian star-names at the Adler Planetarium in Chicago. This was back in the 1970s and 1980s, and today we have, of course, much more sophisticated tools to use at our own convenience, even at our own desktops. We also have much new ancient data to work with, including our uranology texts. A first step could be to test the information embedded in the uranology group against what can be learned from all the other available cuneiform sources. This would leave us with an ancient Mesopotamian star-map consisting of a certain amount of secure identifications, unidentified features, and a set of problems—for example, ancient star-names for which we have no apparent star, and stars for which we have no apparent name. Of course, it is very likely that on occasion we may have two or more different cuneiform names in different sources for what appear to be the same star, and similar apparent anomalies. Further progress might be made by comparing our emerging Mesopotamian star-map with classical maps based on early Greek astronomical works such as the *Phaenomena* of Aratus and other works that can be demonstrated to have their roots in the time before Alexander's conquest of Babylon. These might serve as a chronological benchmark for the opening of regular communication between the Greek and Babylonian scholarship.

For now, however, we present *The Cuneiform Uranology Texts* without further ado. We believe that the presentation of the editions themselves, with our preliminary comments in the Introduction, justifies immediate publication so that the information in our texts can be made available to the widest possible audience. We hope that publication of these materials will allow us, as a community, to move forward together in the widest possible range of directions, with each reader using the texts to his or her own ends and according to his or her own interests and abilities.

APPENDIX A

Star Guide

The Star Guide below lists the stars, constellations, and their parts and elements that are named in the group.[1] The star-names are listed in alphabetical order according to the writing of the name in our texts. Each entry typically gives the Sumerian and Akkadian names of the star, an English translation of the name, and a modern equivalent, followed by a list of occurrences and, when deemed appropriate, some short discussion. Fuller discussion of the stars and their parts is available in the commentaries to the texts belonging to the group.

^{mul}AL.LUL, ^{mul}ALLA = *aluttu*, "The Crab" (Cancer)

ABC Section IV: four-sided figure (*apsamakku*), head
D ii 18–20 (Section F): *apsamakku,* head
D ii 22 (Section G)

AN.GUB.BA.MEŠ[2] (*angubbû*), "The Standing Stars"

ABC Section IX: stars in "The Dog"

AN.TUŠ.A.MEŠ[3] (*antušû*), "The Sitting Stars"

ABC Section IX: stars in "The Dog"

^{mul}*Anunītu*, "Anunitu" (The Eastern Fish in Pisces)

D i 6–7: tail, 9: stretched-out neck (Section A)

> This constellation, together with "The Swallow" (^{mul}sim-mah), forms the classical constellation Pisces. In D i 8–9, Anunitu and "The Swallow" are connected at their tails by ^{mul}ZIB, "The Tails."

[^{mul}ANŠE.KUR.RA = *sīsû*], "The Horse" (Pegasus and Cassiopeia)

D i 25 (Section B), restored

[1] Only entries that must be fully restored are indicated by brackets.
[2] MUL.MEŠ AN.GUB.BA.MEŠ.
[3] MUL.MEŠ AN.TUŠ.A.MEŠ.

Bīt sakkî

ABC Section IX: six stars in "The Dog"

^{mul}DAR.LUGAL = *tarlugallu*, "The Rooster" (Lepus)

D ii 17 (Section E)

^{mul}DILI.BAD, Venus[4]

D i 9, 11 (Section A)
D iv 8' (Section O)

$^{mul}E_4.RU_6$, Eru

ABC Section VI: human figure, clothed, set with a *kurkurru*, head, right hand (whip, lash of the whip), left hand (star)

> This is another clothed divine-human figure, but female and so without a beard.

^{mul}EN.TE.NA.BAR.HUM = *Habaṣīrānu*, "The Mousy One" (Centaurus)

D iii 7–9 (Section K): human figure, head, right hand[5]

> What is left of our text describes this constellation as a human figure with one star for its head, and makes mention of one of its hands.

^{mul}GÀM = *gamlu*, "The Crook" (Auriga)

D iii 16 (Section L)
E 7'–11' (Section I): head of a ram, handle, shaft

^{mul}GENNA = *kajjamānu*, "Saturn"

D iv 8' (Section o)

$^{mul}GU_4$.AN.NA = *is lê*, "The Bull of Heaven" = "The Jaw of the Bull" (Taurus)

E 15' (Section II): bull

> For "The Bull" drawn on VAT 7851, see plate 1.

$^{mul}GU_4$.UTU = *šihṭu*, Mercury

D i 11 (Section A)
D iv 8' (Section O)

> For Mercury drawn between "The Raven" and "The Furrow" on VAT 7847 + AO 6448, see plate 1.

[4] For Akkadian equivalents see, e.g., Kurtik (2007, 108).
[5] See also E 16'.

$^{mul\ d}$*Gula*, Stellar Gula

ABC Section IX–X: divine-human figure, clothed, [head], right hand (star), left hand, chair with legs

> The constellation is the Goddess Gula in the sky. Section IX explains the location of "The Dog" in reference to Stellar Gula, and the constellation is then described in Section X.

^{mul}IBILA.<É>.MAH, "The Heir of the Sublime <Temple>"

D iv 15 (Section P)

> For a discussion of this star with an attempted identification, see the commentary to D iv 15, p. 60.

^{mul}IKU = *iku*, "The Field" (Pegasus)

D i 1–4, 6, 10 (Section A)

^{mul}*immertu* (^{mul}U$_8$), "The Ewe"

D iii 20–21 (Section L)

^{mul}KAK.PAN = "The Arrow" (Sirius)

D ii 21–22 (Section G)

> "The Arrow" has a left foot and elbow in *The GU Text* (Hunger and Pingree 1999, 91), suggesting that this constellation had a version in human form.[6]

^{mul}KA$_5$.A = *šēlebu*, "The Fox"

D iii 18–19 (Section L)

^{mul}KU$_6$ = *nūnu*, "The Fish" (Piscis Austrinus)

D iv 7' (Section O)

^{mul}LÚ.HUN.GÁ = *agru*, "The Hired Man" (Aries)

D i 12–16 (Section A): lamb, ram, forehead, thigh, feet

> Section A offers exegesis justifying how the Babylonian "Hired Man," the constellation in human form (Sumerian lú), could be the same set of stars as the classical ram figure (Sumerian LU, read udu).

^{mul}MAR.GÍD.DA = *eriqqu*, "The Wagon" (The Big Dipper)

ABC Section VII: ***apsamakku*** (wagon) pole

[6] For constellations in both human and other forms, cf. the discussion of "Stellar Gula" in ABC Sections IX–X (see pp. 32–33), and Pingree and Walker (1988, 316, no. 1), with reference to the issue of "The Bow" and Babylonian views of Virgo.

D iii 13–18, 20 (Section L): wagon, *apsamakku*, [pole], cart-pole[7]

The old English name "Charle's Wain" also perceives "The Big Dipper" as a wagon.

mulMAR.GÍD.DA.AN.NA = *eriqqi šamê/šamāmi*, "The Wagon of Heaven" (The Little Dipper, Ursa Minor)

ABC Section VIII: wagon box, pole

mulMAŠ.TAB.BA.GAL.GAL = *tū'amū rabûtu*, "The Great Twins" (Gemini, α and β Geminorum)

ABC Section II: two human figures, bearded, set with a *kurkurru*, heads, hands

Front figure: *hinšu* in his left hand

Back figure: (crescent-shaped) sickle-axe, in his left hand

D ii 1–3 (Section C):[8] front figure: *hinšu* : *ku-ut-ta-ú* in his right hand, whip, lightning bolt in his left hand;
Back figure: sickle-axe in his right hand, whip in his left hand

D iv 16' (Section P)

The hands and feet of the front figure of "The Great Twins" are also noted in *The GU Text*.[9]

mulMAŠ.TAB.BA.TUR.TUR = *tū'amū ṣehrūtu*, "The Lesser Twins" (Gemini)[10]

ABC Section III: two human figures, clothed, bearded, set with a *kurkurru*, heads, hands

Front figure: whip and lash in his right hand

Back figure: lightning bolts in both hands

D ii 4–9 (Section D): two human figures, clothed, bearded, set with a *kurkurru*, heads, hands

Front figure: whip and lash in his right hand

Back figure: lightning bolt in his right hand

D iv 18' (Section P)

mulMAŠ.TAB.BA *šá ina pāni* mulSIPA.ZI.AN.NA, "The Twins Which Stand in Front of 'The True Shepherd of Anu'"[11]

D ii 12–16 (Section E): two human figures, clothed, carrying a large jug in their right hands.

Front figure: bearded

Back figure: face of Latarak

---

7 [*mašaddu*], *zar*[*û*].

8 D ii 1–3 (Section C) gives the end of a section for "The Great Twins" in which the star-name no longer survives.

9 Hunger and Pingree (1999, 91).

10 Ibid., 276. We suggest ζ and λ *Geminorum*.

11 Ibid., 276: γ *Geminorum*.

D iv 18'–20' (Section P)

[mul.mul], "The Stars" (Pleiades) (restored)

E 14' (Section II, restored)

^{mul}MUŠ = *nirahu*, "The Snake" (Hydra)

D iii 1–3 (Section I): wings, feet, a raven on its tail

^{mul}*Ninmah*, Ninmah (Vela)

D iii 4–6 (Section J): clothed, set with a *kurkurru*, right and left hands

[^{mul}nu.muš.da] = *namaššû* (A part of Centaurus, restored)

D iii 12 (Section K, restored)

^{mul}PAN = *qaštu*, "The Bow" (The back part of Canis Major)

D ii 23–24 (Section H)

^{mul}SAG.ME.GAR (Jupiter)

ABC Section IV
D i 11 (Section A)
D ii 20 (Section F, restored)
D iv 8' (Section O)

> For Jupiter drawn between "The Lion" and "The Crab" on VAT 7847 + AO 6448, see plate 1.

^{mul}SIM.MAH = *šinūnūtu*, "The Swallow" (The western fish of Pisces and western part of Pegasus)

D i 4–5 (Section A), bird, winged, flying, 7, tail, 9

> Together, "The Swallow" and Anunitu compose the classical constellation Pisces. See also ^{mul}*Anunītu*.

^{mul}SIPA.ZI.AN.NA = *Šitadallu*,[12] "The True Shepherd of Anu" (Orion)

D ii 9 (Section D)
D ii 10–12 (Section E): human figure, bearded, set with a *kurkurru*, lock and key
D ii 13, 17 (Section E)
D iv 19' (Section P)
E 16' (Section II)

> The right hand and rear heel of the constellation are noted in *The GU Text*. The Sumerian name allows for alternate translations "The True Shepherd of Anu" and "The True Shepherd of Heaven," because Sumerian is used for both An, the King of the Heavens, and his realm (an = *šamû*). We prefer the former because the associations between the constellation and Anu's viziers suggest that the constellation was seen as a servant of Anu. The Akkadian name, and its variations, come from a different tradition. This name is

[12] Also *šitadaru*, *tišattalu*, and *šidallu*.

interpreted in GSL 163–164 and *Urgud* (MSL XI 41 44) as *ša ina kakki mahṣu*, "the one who is smitten with the weapon."[13]

^{mul}*ṣalbātānu*, Mars

D iv 9' (Section O)

^{mul}ŠAH = *šahû*, "The Pig" (Delphinus)

D i 24 (Section B)

^{mul}ŠU.GI = *šību*, "The Old Man" (Perseus)

ABC Section I: clothed, bearded, lash in right hand, left hand, head
E 1'–6' (Section I, restored): human figure, clothed, bearded, set with a *kurkurru*, right hand (crook)

Šullat and Haniš (two stars in Centaurus$^?$)[14]

D iii 10–11 (Section K)

^{mul}UD.KA.DUH.A, "The Demon with The Gaping Mouth" (Cygnus and part of Cepheus)

D i 17–23 (Section B): human figure, clothed, wearing a tiara, two faces—front face in human form, back face in lion form, bearded, open mouth—winged, right shin, right heel, and *lower extremities* (KI.ÚR) in human form

The Akkadian version of this name varies. 5R 46 no. 1 rev. 43 gives a literal translation: ^{mul}ud.ka.duh.a = *ūmu nā'iru*, "Raging Storm/Demon." CAD N_{II} 234–235 and Kurtik (2007, 520) give the Akkadian name as *nimru*, "The Panther," based on a suggestion by S. Parpola in LAS 2 93. However, as CAD N_{II} itself notes: "the reading of ^{mul}UD.KA.DUH.A as *nimru* is not attested." A loan-word, *kuduhhu*, occurs in the Boghazkoi Prayer to the Gods of the Night, KUB IV 47 rev. 46 (^{mul}*ka$_4$-ad-du-uh-hu*).

^{mul}UR.GI$_7$ = *kalbu*, "The Dog" (Hercules)

ABC Section IX: dog, [face], chest, tail

^{mul}UR.GU.LA, ^{mul}UR.MAH, *urgulû*,[15] "The Lion" (Leo)

ABC Section IV: (^{mul}UR.GU.LA)
ABC Section V: tail, *huruppu*, chest (^{mul}UR.MAH)

For a drawing of "The Lion," labeled ^{mul}UR.GU.LA, on VAT 7847 + AO 6448, see plate 1

^{mul}ZIB, *zibbātu* "The Tails" (Pisces)

D Section A, i 8

---

[13] For a more detailed discussion of the Akkadian names, see Horowitz Alb (68).
[14] Kurtik (2007, 202, 508).
[15] *Urra* XXII 291' gives variants ^{mul}ur-gu-la and ^{mul}ur-mah in the Sumerian column, equating the two with the name of the god Latarak in the Akkadian column.

CONSTELLATIONS PARTS AND ELEMENTS[16]

agû, "tiara"

D i 18 (Section B): "The Demon with The Open Mouth"

apsamakku, "Four Sided Figure"

ABC Section IV: "The Crab"
ABC Section VII: "The Wagon"*
D ii 18 (Section F): "The Crab"*
D iii 14 (Section L): "The Wagon"

asīdu, "heel"

D i 22 (Section B): "The Demon with The Open Mouth"

birqu, "lightning-bolt"

ABC Section III: back figure of "The Lesser Twins"
D ii 2 (Section C): front figure of "The Great Twins"
D ii 9 (Section D): back figure of "The Lesser Twins"

dirratu, "lash"

ABC Section I: "The Old Man"
ABC Section III: front figure of "The Lesser Twins"[17]
ABC Section VI: Eru
D ii 8 (Section D): back figure of "The Lesser Twins"

In ABC Section VI, the lash is part of the whip (*qinnazu*). In ABC Section III and D ii 8, the lash is held together with a whip (*iltuhhu*).

gamlu, "crook" (part of "The Old Man")

E 6'–7' (Section I): ^{mul}ŠU$^{?}$.GI$^{?}$ (followed by a descriptions of the constellation "The Crook")

GÚ.LÁ, "stretched-out neck"

D i 9 (Section A): Anunitu

hinšu, "*goad*$^{?}$"

ABC Section II: front figure of "The Great Twins"
D ii 1 (Section C): front figure of "The Great Twins"*

huruppu

ABC Section V: "The Lion"

[16] In the case of multiple occurrences, asterisks (*) indicate entries for which there is detailed discussion of the constellation part or element in the commentaries.
[17] Perhaps also ABC Section I C 3′ in a broken context; see p. 27.

idu, "hand"

passim

iltuhhu, "whip"

ABC Section III: "The Lesser Twins," front figure*
D ii 2–3 (Section C): "The Great Twins," front figure, back figure
D ii 7–8 (Section D): "The Lesser Twins," front figure

irtu, "chest"

ABC Section V: "The Lion"

kablu, leg (of Stellar Gula's chair)

ABC Section X

kalakku, "wagon box"

ABC Section VIII: "The Wagon of Heaven"

kappu, "wing"

D i 5 (Section A): "The Swallow"
D i 21 (Section B): "The Demon with The Open Mouth"
D iii 2 (Section I): "The Snake"

KI.ÚR, "_lower extremities_"

D i 22 (Section B): "The Demon with The Open Mouth"

kinṣu, "shin"

D i 21 (Section B): "The Demon with The Open Mouth"

kurkurru

ABC Section II: "The Great Twins"
ABC Section III: "The Lesser Twins"
ABC Section VI: Eru
D ii 6 (Section D): "The Lesser Twins"
D ii 11 (Section E): "The True Shepherd of Heaven"
D iii 5 (Section J): Ninmah
D iv 5' (Section N): [Star-Name Broken]
E 4'–5' (Section I) "The Old Man"

What *kurkurru* might be in our group is not clear. Here and elsewhere, all known examples of *kukurru* in astronomical contexts belong to constellations in human form, what our group identifies as *ṣalmu*. Yet, not every constellation in human form has a *kurkurru*. CAD K 563–564 gives what may be two relevant homonyms: *kurkurru* A, "a bowl or container"; and *kurkurru* B, "a luminous phenomena?" where CAD lists examples of the word in our group (CAD K 564, f 2'). We suggest that the two words are actually one; that *kurkurru* in our group are luminous phenomena in the form of bowl or container (basically circular in shape) that is placed near the human form of the constellation but not actually held by the human forms in their hands. In our group, *kurkurru* are always listed with a form of the verb *šakānu*, "to set, to place, to position." In contrast, when constellations hold objects in their hands, the verb *ṣabātu* is used.

***kussû*, "chair"**

ABC Section X, Stellar Gula

***ku-ut-ta-ú*, *a large jug*?[18]**

D ii 1 (Section C): front figure of "The Great Twins"*
D ii 16 (Section E): both figures of "The Twins which stand in front of 'The True Shepherd of Anu'"

***mašaddu*, "cart-pole"**

ABC Section VII–VIII: "The Wagon," "The Wagon of Heaven"
D iii 17 (Section L): "The Wagon" (restored)

***namzaqu*, "key"**

D ii 12 (Section E): "The True Shepherd of Anu"

naṣrapu

E 12'–13' (uncertain context)

***panū*, "face"**

ABC Section IX: "The Dog" (restored)
D i 19–20 (Section B): "The Demon with The Open Mouth"
D ii 15 (Section E): "The Twins" in front of "The True Shepherd of Anu" (back figure, face of Latarak)

***pāšu*, "sickle-axe"**

ABC Section II: back figure of "The Great Twins"*
D ii 3 (Section C): back figure of "The Great Twins"

***pû*, "mouth"**

D i 21 (Section B): "The Demon with The Open Mouth

***pūtu*, "forehead"**

D i 15 (Section A): "The Hired Man"

***qinnazu*, "whip"**

ABC Section VI: Eru

[18] See the discussion, p. 49.

***quppû*, knife**

D iv 6' (Section N)

***rapaštu*, "thigh"**

D i 16 (Section A): "The Hired Man"/classical Aries

***rēšu*, head**

passim

> A single star typically represents the heads of constellations in both human and animal form; for example, "The Crab" in ABC Section IV. Thus, two stars represent the two heads of twins constellations; for example, "The Great Twins" and "The Lesser Twins" in ABC Sections II–III and D ii 6–7 (Section D).

***ṣipri*, "crest"**

D i 8 (Section A), "The Tails"

***šašallu*, "heel"**

ABC Section VII: Eru

***šēpu*, "foot"**

D i 16 (Section A): "The Hired Man"/classical Aries
D iii 2 (Section L): "The Snake"

***uppu*, "lock"**

D ii 12 (Section E): "The True Shepherd of Anu"

***ūru*, "shaft" (of a crook)**

E 11' (Section I): "The Crook"

***uskaru*, "crescent"**

ABC Section II: back figure of "The Great Twins" (describing the shape of the sickle-axe)

***zibbatu*, "tail"**

ABC Section V–VI: "The Lion"
ABC Section IX: "The Dog"
D i 7 (Section A): "The Swallow" and Anunitu
D iii 3 (Section I): "The Snake"

***ziqnu*, "beard"**

passim

APPENDIX B

BM 66958 Side B

BM 66958
Photograph: plate 7
Copy: plate 9

Edition

1' ⸢DIŠ⸣ [*ina*] ⸢^{itu}APIN$^{?}$ UD.10⸣.KAM [. . .]
2' EGIR.MEŠ *šá* MUL P[A$^{?}$ . . .]
3' (blank) ⸢x⸣ [. . .]

4' DIŠ *ina* ^{itu}GAN UD.3.KAM *ina* ^{d}UTU.È ⸢x⸣ [. . .]
5' IGI-*it* ^{d}*li*$_{9}$-*si*$_{4}$ 10 UŠ [. . .]

6' DIŠ ^{mul}UDU.IDIM.GU$_{4}$.UD UD.16$^{?}$.[KAM . . .]

7' DIŠ *ina* ^{itu}BÁR UD.28.KAM *ina* ^{d}U[TU.ŠÚ . . .]
8' DU-*ma* ŠÚ 1 ITU ⸢GU$_{4}$$^{?}$⸣ [. . .]

9' DIŠ *ina* ^{itu}ŠU UD. ⸢12⸣.K[AM . . .]
10' UD.26.KA[M . . .]

11' ⸢DIŠ *ina* ^{itu}.KIN UD⸣.[x.KAM . . .]

Translation

1' ¶ Month VIII$^{?}$, 10th day [. . .]
2' Rear (Stars) of P[abilsag$^{?}$ . . .]
3' (blank) . . . [. . .]

4' ¶ Month IX, 3rd day, in the east [. . .]
5' in front of Lisi 10° [. . .]

6' ¶ Mercury, 16$^{?}$[th] day, [. . .]

7' ¶ Month I, 28th day, in the [west . . .]
8' it moves forward and has its last visibility. . . . Month I[I$^?$ . . .]

9' ¶ Month IV, 12t[h] day [. . .]
10' 26t[h] day [. . .]

11' ¶ Month VI, [xth] day [. . .]

Commentary

The text contains a collection of observations of the first and last visibilities as an evening and morning star of an inner planet. With the exception of line 6', the text is divided into sections, each of which contains the dates of a pair of observations of the first and last visibility either as an evening star or a morning star, sometimes accompanied by a reference to the position of the planet, written over 2 or 3 lines. Line 6' comprises a section of its own and begins with the name of the planet Mercury (written ^{mul}UDU.IDIM.GU$_4$.UD) followed by a day number (but no month name); it is unlikely that this line would contain enough space for the dates of both first and last visibilities. With this in mind, it needs to be considered whether the sections before and after line 6' are part of the same set of observations or whether they refer to different planets.

The data in lines 1'–5' are summarized below:

Month VIII$^?$ 10	First visibility in the [west]
[Month x x]	Last visibility in the [west]
Month IX 3	First visibility in the east
[Month x x]	Last visibility in the east

This pattern of the dates of first and last visibilities and the lengths of the periods of visibility and invisibility they imply are characteristic of Mercury and no other planet. This is further confirmed by the stars mentioned in lines 2' and 5'. The name of the stars in line 2' can be restored as "The Rear (Stars) of [Pabilsag]," a group of stars named in later texts 4-ÀM *ár šá* PA "The Four Rear (stars) of Pabilsag,"[1] which have a celestial longitude of about 260 degrees. In line 5' the planet is said to be "in front of Lisi," a star with a longitude of about 238 degrees. Mercury moves retrograde between its last visibility in the west and its first visibility in the east, in agreement with the positions stated in the text.

Following line 6', we have another run of the dates of first and last visibilities of Mercury:

Month I 28	First visibility in the [west]
Month II$^?$ [x]	Last visibility in the [west]
Month IV 12	First visibility in the [east]
Month IV 26	Last visibility in the [east]
Month VI$^?$ [x]	First visibility in the [west]

---

[1] Roughton and Canzoneri (1992).

Once more, the preserved dates are in the right range for Mercury. Thus, both before and after line 6', we are dealing with Mercury data. However, it does not appear that the data before and after line 6' follow each other in consecutive years; in particular, there should be another pair of first and last visibilities in either or both the evening and the morning after month IX and before month I of the following year. For this and other reasons, it is not possible to date the observations in the text.

It is worth noting that the terminology used in the text is suggestive of a date in the Neo-Assyrian or early Neo-Babylonian period. Note, for example, the use of ^{d}UTU.È and ^{d}UTU.[ŠÚ] rather than NIM and KUR for east and west in the context of visibility phenomena; MUL rather than the later MÚL, ^{d}*li*$_{9}$-*si*$_{4}$ rather than SI$_{4}$ for the star Lisi (α Scorpii); and the references to stars in the reports of observations of first visibility, which is very rare in later texts.

APPENDIX C

The Late Babylonian Fragment MLC 1884

INTRODUCTION

MLC 1884 is a fragmentary Late Babylonian clay tablet from the Yale Babylonian Collection, originating from the Morgan Library Collection assembled by Albert T. Clay. Nothing specific is known about the circumstances of its discovery, but considering that many MLC tablets can be shown to come from Uruk, and taking into account that the obverse of MLC 1884 seems to refer to the topography of that city, the tablet was almost certainly found at Uruk as well.

The tablet is written in a script that is characterized by vertical wedges heavily tilted to the left, which, together with the damaged surface of the obverse, makes it hard to read. The distinctive ductus, also known from other scholarly texts from Hellenistic Uruk (e.g., TU 20, 23, 24, 26, 29, 35 [colophon]), suggests that the tablet is quite late. The sign forms bear a strong resemblance with the forms attested on a small group of archival tablets from Uruk from the Hellenistic period, now mostly housed at Yale, that were published and/or discussed by Beaulieu (1989). One of them, MLC 1853, is, like our text, part of the Morgan Library Collection. Four of the tablets examined by Beaulieu are dated, to 253–252, 244, 240, and 202 B.C.E., respectively, and it is possible that MLC 1884 was written at about the same time, probably by a scribe closely affiliated with the (*Bīt*) *rēš* temple.[1] A later date, however, is not excluded.[2] The latest dated archival text from Uruk is from 108 B.C.E. (Kessler 1984), and the (*Bīt*) *rēš* survived even longer; it seems to have been destroyed by a fire shortly after 90 B.C.E. (Kose 1998, 51–2, 133).[3]

MLC 1884 is included in this book because one of its sides contains descriptions of how to draw celestial constellations. Because there seems to be no space

[1] Beaulieu (1989, 60) points out that most of these archival documents display an "écriture inclinée." This sets them apart from the legal documents from Hellenistic Uruk, which are written in "regular" script.
[2] For photos of cuneiform signs on tablets from Babylon and Uruk from the Late Achaemenid, Seleucid, and Parthian periods, see Ossendrijver (2017).
[3] Hunger and de Jong (2014) argued that an astronomical Almanac found in the south-eastern area of Uruk might date to A.D. 79/80, which would make it the latest datable cuneiform tablet known thus far. Owing to the fragmentary nature of the text, this dating remains, however, somewhat uncertain. The tablet is likewise written with tilted vertical wedges.

for a colophon where the text breaks off,[4] the side in question is more likely the reverse than the obverse, even though some uncertainty remains.[5] The celestial section of MLC 1884 is presented below in transliteration only; a translation and commentary are found in Chapter 4.

Unlike the "reverse," the poorly preserved presumed obverse of the tablet deals not with the heavens, but with the earth: it describes topographic features, both temples and watercourses, apparently found in (and around?) the city of Uruk. This juxtaposition of the celestial with the terrestrial sphere is what sets MLC 1884 apart from the other tablets edited in this book and makes it quite interesting. Unfortunately, deciphering what is left of the—apparently unique—topographic notes is quite difficult. One can only hope that future studies will eventually solve some of the remaining problems.

Transliteration

obverse? (several lines missing)

1'. [(x)] ⌜x⌝ ⌜ki$^{?}$ x x x x x x⌝
2'. [o] [(x)] ⌜$^{d?}$*a*$^{?}$⌝-*nun-ni-t*[*u_4*]
3'. [AŠ-i]ku 20 ⌜SAR⌝ [x (x)] ⌜x⌝ ⌜É⌝ ^{d}[x x]
4'. [AŠ]-iku 10 SAR ⌜x x x x (x) É$^{?}$⌝ [x x]
5'. AŠ-iku 10 SAR ⌜é$^{??}$ x⌝ na$^{?}$ x ⌜É⌝ ^{d}⌜x x (x)⌝
6'. AŠ-⌜iku⌝ 20 ⌜SAR é$^{??}$⌝ ⌜x⌝ ni ni ⌜É⌝ ^{d}⌜x x (x)⌝
7'. AŠ-iku ⌜É⌝ ^{d}⌜x⌝
8'. AŠ-iku ⌜É⌝ ^{d}IGI.⌜DU⌝
9'. [(n+)] 40 ⌜x x⌝ aš ⌜x⌝
10'. [x] DIŠ+KÙŠ$^{??}$ ⌜x x⌝ za$^{??}$ DINGIR.MEŠ ⌜x x x x x ki$^{?}$⌝

11'. A.⌜ENGUR$^{?}$ MAḪ$^{?}$ 20$^{?}$ x PAB$^{?}$-E$^{?}$⌝ 10 E
12'. ⌜40⌝(+n) *miṭ-ra-tu_4* *ma-a-tú*
13'. TA EDIN *a-tap-pi* EN *muḫ-ḫi* É-*re-eš*
14'. *a-tap-pi* MU-*šú*
15'. *a-na mé-*⌜*eḫ*$^{?}$⌝*-ret* É.AN.NA DIM_4(PAB-PAB)-*ma* PA_5(PAB-E) MU-*šú*
16'. ⌜*ke*⌝*-er-ḫi* KUR-*ma* E MU-*šú*
17'. [x] ÍD *ki-sur-re-e* MU-*šú*
18'. [(x)] x È(UD.DU)-*ma ú*$^{?}$*-ra-šú* MU-*šú*
19'. [(x)] x uš *a-tap-pi*
20'. [x x x x] ⌜x x x x⌝ [x]

(one or two lines lost?)

reverse?

1. [x x] ⌜x x⌝ [x x x x x x]
2. [o] ⌜x (x)⌝ [x (x)]

---

[4] This assessment is based on the curvature of the tablet and the observation that the part where the "uranographic" description breaks off is significantly thinner than the preceding portion.

[5] It is possible that the tablet did not have a colophon, or that the colophon did not comprise more than one line. In this case, the side with the "uranographic" description could also represent the obverse.

3. ⸢*ṣa*⸣-*lam šu-ú lu-bu-uš-tu*$_4$ *la-b*[*i-iš*]
4. *ziq-nu za-qin-nu ku-ur-ku-ru šá-kin*
5. DIŠ+*en* MUL *ina ku-ur-ku-ru-šú* GUB-*zu*
6. *ina* ŠU^{II} 15-*šú* GÀM *na-ši*
7. *ina*$^{?}$(erased?) GÀM DIŠ+*en* ^{mul}GÀM GUB-*zu*
8. SAG.DU GÀM SAG.DU UDU.NÍTA
9. ⸢2(or 3)⸣ MUL *ina* SAG.DU GÀM *e-ṣir*.MEŠ
10. [n M]UL *ina rit-ti* GÀM GUB-*zu*
11. [M]UL SA$_4$ (or [^{m}]^{ul}SA$_4$) *ina ur-šú* GUB-*zu*
12. [x]⸢(x) x⸣-*šú*$^{?}$ ^{d}NE$^{?}$.GI$^{?}$ *na*$^{?}$-*áš*$^{?}$-*rap na-ši*
13. [x x x] ⸢x x⸣ *ina*$^{?}$ *na*$^{?}$-*áš*$^{??}$-*rap e-ṣir*.MEŠ

---

14. [MUL$^{?}$.MUL$^{?}$] ⸢$^{d?}$IMIN$^{?}$.BI$^{?}$⸣ DINGIR.MEŠ GAL.MEŠ
15. [GU$_4$$^{?}$.AN$^{?}$.N]A$^{?}$ ⸢*is*$^{?}$ *le*$^{?}$-*e*$^{?}$⸣ *a-lu-ú*
16. [x] ⸢x x (x)⸣[E]N$^{?}$.TE.NA.BAR.ḪUM$^{?}$
17. [x] *ina*$^{?}$ *šap*$^{?}$-*li*$^{?}$ ⸢x x (x) $^{mul?}$⸣*nu-mu*[*š-da*]
18. [x (x) *a*$^{?}$]-*ge*$^{?}$-*e* ⸢DINGIR ŠEŠ$^{?}$.MEŠ⸣ *ki-lal-la-n*[*u*$^{?}$](or: *a*[*n*$^{?}$])
19. ⸢x x⸣ *pa-ni* UR.MAḪ *la-an-nu šá ṣa-la*[*m*]
20. ⸢x x x x x x x x x⸣ [x]

remainder lost

Translation (Obverse Only)

1'–2' . . . . . . (the goddess) Anunnītu;
3' one *ikû* and 20 *mušarû* [. . .] . . . the temple (*bītu*) of [. . .];
4' one *ikû* and 10 *mušarû* . . . the temple of [. . .];
5' one *ikû* and 10 *mušarû* . . . the temple of . . . ;
6' one *ikû* and 20 *mušarû* . . . the temple of . . . ;
7' one *ikû*, the temple of . . . ;
8' one *ikû*, the temple of (the god) IGI.DU (= Palil?);
9'–10' [. . .] 40(?) . . . [. . .] cubit(s)(??) . . . the gods . . .

11'–12' The "Exalted River(?)," 20(?) . . . *palgu*-canals(?), 10(?) *iku*-ditches (or: dykes), 40(+n) *miṭirtu*-watercourses.
12'–13' The area (lit., "land") from the terrain of the *atappu*-canal up to the (*Bīt*-)*rēš*, 14' its name is "*atappu*-canal";
15' (the area that) approaches the Eanna-temple, its name is "*palgu*-canal";
16' (the area that) reaches the *kerḫu*-enclosure wall, its name is "*iku*-ditch (or: dyke)";
17' (the area that (?)) [. . .] the river, its name is "*kisurrû*";
18' (the area where the canal (?)) exits / departs from [. . .] . . . , its name is "*urāšu*-plot";
19' [. . .] . . . the *atappu*-canal 20' [. . .] . . . [. . .].

Philological Notes (Obverse Only)

1'. The first sign(s) could be [AŠ]-⸢iku⸣, as in lines 3'–8', but what follows is clearly neither a number nor É.
2'. Whether the broken space before the divine name is large enough for a restoration of the expected [É] ("temple of") is not entirely clear, but it

probably is not. The reading ⸢^{d}*a*⸣- is uncertain as well. Note that Anun(n)ītu is attested as a part of the constellation Pisces in the "uranographic" tradition (Text D, I 6–7, 9).

3'. One *ikû* corresponds to ca. 3,600 m^2 or 100 *muš/sarû*(SAR); one *muš/sarû* is, hence, ca. 36 m^2.

4'. The sign after SAR could be ⸢É⸣. Just as in the following two lines, this ⸢É⸣ might introduce a ceremonial temple name in Sumerian.

7'. Neither DINGIR.MAḪ nor ^{d}IŠKUR seems to fit the traces.[6]

9'. It is tempting to read at the end of the line AŠ-⸢iku⸣, but there is a notable gap between the two signs, unlike in the other cases in which AŠ-iku is attested in the text.

11'. The signs at the beginning of the line are not quite clear, but a reading A.⸢ENGUR MAḪ⸣ seems to fit the traces better than A.⸢ENGUR LUGAL⸣ (note the MAḪ in rev. 19). The first two signs are, moreover, rather not A.⸢ŠÀ⸣.[7] Admittedly, an "Exalted River" seems so far unattested in Late Babylonian texts from Uruk,[8] whereas the *Nār šarri*, the "Royal River," is known to have played an important role in the city's landscape during the first millennium B.C.E. However, as argued below (see "General observations"), there are reasons why we should not expect the *Nār šarri* to occur in our text.

It would seem that obv. 11'–12' provide numbers for the various types of smaller canals diverted from the main river. The reading ⸢PAB-E⸣ (= *iku*) is not without problems, because the sequence *palgu* – *iku* is unusual—one would rather expect *iku* – *palgu*. Note, however, that *palgu* seems to precede *iku* in lines 15'–16' as well. For a discussion of the meanings of *iku*, *palgu*, and also *atappu* (obv. 13', 14', 19'), based primarily on Assyrian texts, see Bagg (2000, 142–5). The terms designate various types of small canals but are apparently not used consistently. Instead of "10," one could, in principle, also read *u* "and."

12'. For the reading *miṭ-ra-tu*$_4$ instead of *be-ra-tu*$_4$ or *mid-ra-tu*$_4$, see CAD M/2 144–5. The term designates, on one hand, a specific type of field and, on the other, a type of canal or ditch (Bagg 2000, 39); here, the latter meaning seems to be called for. Because elsewhere on the obverse, lines tend to be identical with semantic units, one is inclined at first glance to consider *mātu* to somehow summarize the preceding items, and this may in fact be the case; but it also makes sense to assume that the word introduces the following section, which seems to name various areas within Uruk through which specific watercourses were flowing.

15'. DIM$_4$ = *isanniq* seems grammatically slightly odd, but it is consistent with the (logographically written) verb forms in the following lines. In fact, the verb *sanāqu*, usually constructed with *ana*, is attested not only with persons but also with boats, goods, canals, and other subjects (see CAD S 135–7). The passage provides important confirmation that the Eanna complex still existed (albeit presumably on a much reduced scale) during the later Achaemenid and Hellenistic periods (see "General Observations").

[6] *Bīt-Adad* was one of the districts of Uruk during the Seleucid period; see Falkenstein (1941, 51).
[7] The reading A.⸢ENGUR⸣ seems preferable to A.⸢ŠÀ⸣ both from an epigraphic point of view and considering the context.
[8] In a Sumerian hymn to Lugalgirra and Meslamtaea, the underworld river is characterized as an "íd maḫ" (Horowitz 2011, 357–8), but this is of little relevance for our discussion.

16'. At first glance, it would seem that the enclosure wall (*kerḫu*) invoked here belonged to the Eanna temple mentioned in the previous line. Apart from two references in a Neo-Babylonian building inscription of an unidentified Late Assyrian(?) king related to a sanctuary of the god Erragal (YOS 9, no. 80), the only other first millennium Babylonian text mentioning the *kerḫu* of a sanctuary is an inscription of Sargon II (RIMB 2, 6.22.4) in which the king claims that he built "the outer enclosure wall, the courtyard of Eanna, the narrow gate, and the regular gate" (*ker-ḫu ki-da-a-nu* KISAL *é-an-na* KÁ *qá-tan u* KÁ *ki-i-nu*). Identifying the *kerḫu* mentioned in MLC 1884 as that of the Eanna faces a serious problem, however: archaeological and philological evidence seems to suggest that much of the earlier Eanna complex was abandoned to ruin since the Late Achaemenid period (see below). It therefore has to be considered that the enclosure wall of another sanctuary is meant here—or perhaps even the so-called "Seleucid Wall" (see Baker 2014, 199, with earlier literature).

17'. The line is syntactically different from the preceding two and the following one, which all include a finite verb. A translation "river of the *kisurrû*" would be grammatically possible as well but seems less likely in this context. Whether the river alluded to is identical with the one mentioned in line 11'(?) or located close to the Eanna is not entirely clear. The word *kisurrû* normally means either "boundary" or, more generally, "territory," but in a bilingual Late Babylonian liturgical text (SBH, p. 49, rev. 14–15, see CAD K, 434a) the word is juxtaposed with *miṭrātu* (see the note above on line 12'), and in the line under discussion here, *kisurrû* most likely serves as a close semantic equivalent of that word as well.

18'. At the beginning of the line, some topographical feature must have been mentioned. Together with *miṭirtu* (see obv. 12') and *ṣippatu*, the term *urāšu* is named in *Malku* II 117–118a as a synonym of *kirû* "garden" (Hrůša 2010, 60–1). Our tablet seems to provide the first attestation of the word in a text from first millennium Uruk.

General Observations (Obverse Only)

The topographic notes found on the obverse of MLC 1884 describe temples and watercourses in the city of Uruk. Most likely, these notes reflect topographical realities of the Hellenistic period, when the tablet was apparently written. What mitigates against the assumption that the notes were copied from an earlier tablet is, first, that there is no duplicate, and second, that the (*Bīt*) *Rēš* temple, mentioned in obv. 13', is so far only attested in Hellenistic texts—and cannot have been a sanctuary important enough to serve as a landmark before the Late Achaemenid period.[9] The topography of Uruk changed substantially between the early Achaemenid and the Hellenistic era: The northern part of the city, for example, apparently largely unoccupied during the former era, saw a substantial increase in settlement during the latter (Baker 2014).

[9] Note, however, that Sennacherib mentions in one of his inscriptions the deportation, in 693 B.C.E., of the statue of a goddess called *Bēltu-ša-rēši* ("Lady of the *Rēš*"), who was probably the resident deity of the sanctuary at that time (RINAP 3/1, 223, line 31). It is hence likely that a small *Rēš* temple already existed in the seventh century B.C.E.

The passage preserved on the obverse opens with a first section (obv. 1'–10') on temples of various deities.[10] Most of the lines begin with references to the areas covered by these temples, either one *ikû* (ca. 3,600 m^2), one *ikû* and 10 *mušarû* (ca. 3,960 m^2), or one *ikû* and 20 *mušarû* (ca. 4,320 m^2). To which specific parts of the sanctuaries the numbers actually refer remains unclear, but considering that the temples listed in the preserved part of MLC 1884 seem to belong to rather minor deities, they seem quite high and probably concern not only the main temple building but also surrounding courtyards and additional edifices.

Another text found at Uruk that uses, inter alia, the linear-based surface units *ikû* and *mušarû* to characterize the dimensions of temples is AO 6555 (TU 32), the famous so-called Esagil tablet (latest edition: George 1992 [BTT], no. 13). AO 6555 was copied from an original from Borsippa in 229 B.C.E. by Anu-bēlšunu, son of Nidinti-Anu, of the Sîn-leqe-unninnī family, for Anu-bēlšunu, son of Anu-balāssu-iqbi, of the Aḫi'ūtu family. The tablet claims that the size of the Grand Court (*kisalmaḫ*) of the Esagil was 1 *ikû* and the Court of Ištar and Zababa ½ *ikû*, and includes, in addition, complex calculations of the ziggurat Etemenanki in Babylon.

Although I am not aware of any other texts using the metrological units *ikû* and *mušarû* in connection with the temples of Uruk, it deserves mentioning that a small tablet found in the area of the (*Bīt-*)*rēš* temple in Uruk (van Dijk 1962, 60–1, plate 28b) states that the sides of the *papāḫu* of the (*Bīt-*)*rēš* measured 150 and 260 cubits (*ammatu*), respectively, and those of the *papāḫu* of the Ešgal temple 260 and 260 cubits. Because one cubit corresponds to ca. 0.5 m, this would indicate that the two sanctuaries had a surface of 9,750 and 16,900 m^2, respectively. In fact, the surfaces of the brick cores of the main buildings of the temples measure ca. 3,935 m^2 in the case of the (*Bīt-*)*rēš* and 8,667 m^2 in that of the Ešgal. The text provides, in addition, extremely high numbers in the (capacity-surface–based) "seed system" for the two sanctuaries. Two additional "topographical" texts that deal with the cultic infrastructure of Uruk, BTT nos. 25 and 31 (George 1992), seem not to include any measurements.

The numbers on the obverse of MLC 1884 also remind one of the famous first tablet of the Gilgameš epic, which provides (exaggerated) measurements for the whole city of Uruk and its four main components: [*šār*] *ālu* [*šār*] *kirâtu šār essû pitir bīt Ištar* / [3 *šār*] *u pitir Uruk tamsīḫu* "[One *šār*] is city, [one *šār*] date-grove, one *šār* is clay-pit; half a *šār* the temple of Ištar: / [three *šār*] and a half (ca. 30 square kilometers) is Uruk, (its) measurement" (George 2003, 538–9, lines 22–23). As pointed out by George, the Neo-Babylonian *šār* corresponds to 108 *ikû*.

Most of the divine names associated with the sanctuaries mentioned in MLC 1884 are lost or illegible, but two, Anun(n)ītu (reading uncertain) and ^{d}IGI. DU(= Palil?), are partially preserved. That Anun(n)ītu and ^{d}IGI.DU, who are minor deities, seem to be credited in the text with major sanctuaries is surprising, even though various sources demonstrate that they were, in fact, worshipped in first millennium Uruk. Both deities are mentioned in a letter from the Neo-Assyrian period, which refers to repair work on their statues, plus a statue of the goddess Kurunnītu(?), in a temple workshop in the city (SAA 10, no. 349, obv. 19–23).[11] Anun(n)ītu is also attested in a text from Uruk from the Seleucid period: The "Fête

[10] For a list of temples in Uruk, see Falkenstein (1941, 52).

[11] For the restoration [^{d}KAŠ.D]IN-*i-ti* instead of [^{d}UNUG^{ki}]$^{!!}$-*i-ti*, see Beaulieu (2003, 321, no. 57).

d'Ištar" mentions her, together with numerous other female deities, as accompanying Ištar on her way to the *Akītu*-house (TU 42+ = Lackenbacher 1977, 45; Linssen 2004, 239, 241, rev. 12').[12] ^{d}IGI.DU, apparently identified in Uruk more with Ninurta than with Nergal, is known to have shared a sanctuary in the Eanna complex with Gula during the Neo-Babylonian period, when he also had a temple in the nearby city of Udannu (Beaulieu 2003, 274–5, 282–95). The deity continued to be worshipped in Uruk well into the Seleucid period, as indicated by a reference to him in the aforementioned "Fête d'Ištar" (TU 42+ = Lackenbacher 1977, 40; Linssen 2004, 238, 240, obv. 7') and in the New Year's ritual of Anu (TU 39, obv. 21; see Linssen 2004, 185, 188), which claims that he and the deities Adad, Šala, Sîn, Šamaš, Ninurta, Messagunug, Lugalbanda, and Ninsun "will rise from their sanctuaries (É.MEŠ)," to take up positions toward Anu in the Grand Courtyard (of the (*Bīt-*)*rēš*).

The section on temples ends, after two poorly preserved lines of uncertain content, with a horizontal ruling that is followed by a second section, which deals with various watercourses in the city of Uruk. The existence of such watercourses in Late Babylonian Uruk has been known ever since the publication of Seleucid period ritual texts related to the procession of the god Anu from the (*Bīt-*)*rēš* in the center of Uruk to the *Akītu*-house outside the city, on its eastern side (see Falkenstein 1941, 45–50). BRM 4, no. 7 (Linssen 2003, 209, 211) mentions several times, in lines 13, 16, and 20, the kar-kù-ga ("Pure Quay") and the "embankment" (*arammu*) of the "boat of Anu" (^{giš}má-an-na). Several legal documents (e.g., BRM 2, no. 21, obv. 2–3, and YOS 20, no. 96, obv. 2–5) demonstrate, moreover, that a watercourse known as the *Nār-Ištar* flowed through Uruk during the Hellenistic period.

Heather Baker, in a forthcoming book on the topography of Neo- and Late Babylonian cities, discusses at least 16 named Urukean watercourses (including a small number that can be discarded).[13] A particularly important one, the *Nār-šarri* ("Royal River"), is mentioned in documents from Neo-Babylonian Uruk (see Zadok, RGTC 8, 385, s. v. *Nār-šarri* 4). The name was used for a former branch of the Euphrates that, according to Kose (1998, 15, fig. 5), flowed toward Uruk from the north and then followed the city wall on the eastern side, effectively forming a city moat. Janković (2010, 419–20) has claimed that a branch of the *Nār-šarri*—known as *Nār-šarri ša qereb Uruk*—entered the city from the north, but as kindly pointed out to me by Baker, this may, actually, not have been the case—there is no evidence that the *Nār-šarri* flowed within the city.[14] As already indicated above in the "Philological Notes," it is therefore unlikely that the river mentioned in obv. 11' of our text is the *Nār-šarri*; reading ÍD MAḪ seems preferable, even though this name is not yet attested in other Hellenistic-period texts from Uruk.

There is also material evidence for the existence of a network of canals within the city of Uruk, as already realized by Falkenstein (1941, 48), who referred to

[12] See also Pongratz-Leisten (1994, 142) and Beaulieu (2003, 311).

[13] Personal communication, August 10, 2017. None of them bears one of the generic names listed in our text (*palgu*, *iku*, *atappu*).

[14] In Baker's view, *ša qereb Uruk* is used also to refer to the area immediately outside of the city proper. At Uruk, according to Baker, the terminology relating to the city's limits was used more loosely than elsewhere, probably reflecting the fact that the city wall was by this time long in ruins. The point is dealt with in more detail in Baker's forthcoming book.

excavation work by A. Nöldeke and a number of aerial photographs. In recent years, geomagnetic and geo-archaeological soundings at Uruk (see Becker et al. 2013; Brückner 2013) have further refined this picture. The investigation of ca. 50 hectares of urban terrain has shown beyond doubt that Uruk, like an "Amsterdam of southern Mesopotamia," was crisscrossed by numerous watercourses, some of substantial dimensions, with a breadth of up to 10 meters.

MLC 1884, despite its deplorably poor state of preservation, provides additional confirmation of this new picture of Uruk. It names a number of canals within the city, indicates, albeit in rather vague terms, their orientation with regard to the (*Bīt-*)*rēš* sanctuary and the Eanna temple, and shows, by using an array of technical terms, that the canals in question were of different types and sizes. It seems likely that the watercourses mentioned in obv. 13'–16' had their origins in a large canal located in the northern area of Uruk, from where they flowed southward.

The reference to the Eanna in obv. 15' is of particular interest, since this sanctuary, the central temple in Uruk during the Neo-Babylonian period, had apparently suffered a long period of decay from the reign of the Persian king Xerxes onward (Kessler 2005; Baker 2014, 188–91). In the course of the later Achaemenid era and the Hellenistic period, new temples, the (*Bīt-*)*rēš* and the *Ešgal*, had replaced the Eanna as Uruk's main religious centers. The massive rebuilding of the (*Bīt-*)*rēš* complex during the second half of the third century BCE is documented by inscriptions written in the names of two local dignitaries with both Akkadian and Greek names, Anu-uballiṭ Nikarchos and Anu-uballiṭ Kephalon (Doty 1988). But the Eanna complex was not entirely abandoned: archaeological evidence, and a few references in cultic texts, suggest that ritual activity in what had remained of the Eanna continued, albeit on a more modest scale, during the Hellenistic period (Kessler 2005, 284; Baker 2014, 189).[15] MLC 1884 confirms that the Eanna complex was indeed not forgotten during this time and was still impressive enough to serve as a landmark in a topographic description of Uruk.

[15] See, moreover, Falkenstein (1941, 40–1) and Kose (1998, 257). Kose argues that only the ziggurat of the Eanna and the Karaindaš temple were restored in Seleucid times, but note that TU 38, a text on daily offerings in the Hellenistic temples of Uruk, mentions not only the (*Bīt*) *Rēš* and the Ešgal, but also the Eanna temple (Linssen 2004, 175, 179, rev. 35, 39, 45).

References

Al-Rawi, F., and W. Horowitz. 2001. "Tablets from The Sippar Library IX: A *Ziqpu*-Star Planisphere," *Iraq* 63: 171–81.

Bagg, A. M. 2000. *Assyrische Wasserbauten*. Baghdader Forschungen 24. Mainz: von Zabern.

Baker, H. 2014. "Temple and City in Hellenistic Uruk: Sacred Space and the Transformation of Late Babylonian Society." In *Redefining the Sacred: Religious Architecture and Text in the Near East and Egypt 1000 BC–AD 300*, Contextualizing the Sacred 1, edited by E. Frood and R. Raja, 183–208. Turnhout: Brepols.

Beaulieu, P.-A. 1989. "Textes administratifs inédits d'époque hellénistique provenant des archives du *Bīt Rēš*," *Revue d'Assyriologie* 83: 53–87.

———. 1992. "Antiquarian Theology in Seleucid Uruk," *Acta Sumerologica* 14: 47–75.

———. 1995. "Theological and Philosophical Speculations on the Name of the Goddess Antu," *Orientalia* 64: 187–213.

———. 1999. "The Babylonian Man in the Moon," *Journal of Cuneiform Studies* 51: 91–9.

———. 2003. *The Pantheon of Uruk during the Neo-Babylonian Period*. Cuneiform Monographs 23. Leiden: Brill/Styx.

———. 2010. "The Afterlife of Assyrian Scholarship in Hellenistic Babylonia." In *Gazing on the Deep: Ancient Near Eastern and Other Studies in Honor of Tzvi Abusch*, edited by J. Stackert, B. N. Porter, and D. P. Wright, 1–18. Bethesda: Eisenbrauns.

Becker, H., M. van Ess, and J. Fassbinder. 2013. "Uruk: Urbane Strukturen im Magnet- und Satellitenbild." In *Uruk: 5000 Jahre Megacity*, edited by N. Crüsemann et al., 354–61. Petersberg: Michael Imhof Verlag.

Black, J., and A. Green. 1992. *Gods, Demons, and Symbols of Ancient Mesopotamia: An Illustrated Dictionary*. London: British Museum Press.

Bloch, Y., and W. Horowitz. 2015. "Urra = *hubullu* XXII: The Standard Recension," *Journal of Cuneiform Studies* 67: 71–125.

Britton, J. P. 2002. "Treatments of Annual Phenomena in Cuneiform Sources. In *Under One Sky: Astronomy and Mathematics in the Ancient Near East*, edited by J. M. Steele and A. Imhausen, 21–78. Münster: Ugarit-Verlag.

Brückner, H. 2013. "Uruk – aus geoarchäologischer Sicht." In *Uruk: 5000 Jahre Megacity*, edited by N. Crüsemann et al., 343–51. Petersberg: Michael Imhof Verlag.

Burstein, S. M. 1978. *The Babyloniaca of Berossus*. Malibu: Undena Publications.

Doty, L. T. 1988. "Nikarchos and Kephalon." In *A Scientific Humanist: Studies in Memory of Abraham Sachs*, edited by E. Leichty et al., 95–118. Philadelphia: University Museum.

Falkenstein, A. 1941. *Topographie von Uruk I: Uruk zur Seleukidenzeit*. Ausgrabungen der Deutschen Forschungsgemeinschaft in *Uruk*-Warka 3. Leipzig: Harrassowitz.

Fincke, J., and W. Horowitz. "New Information about the *ziqpu*-Stars, the Sun and Cubits Based on one New Tablet (BM 37373) and Some Previously Known Texts" (forthcoming).

George, A. 1991. "Babylonian Texts from The Folios of Sidney Smith, Part Two," *Revue d'Assyriologie* 85: 137–67.

______. 1992. *Babylonian Topographical Texts*. Orientalia Lovaniensia Analecta 40. Leuven: Department Oriëntalistiek and Uitgeverij Peeters.

______. 2003. *The Babylonian Gilgamesh Epic: Introduction, Critical Edition and Cuneiform Texts*. Oxford: Oxford University Press.

Hallo, W. W. 2008. "Mul.Apin and the Names of the Constellations." In *Studies in Ancient Near Eastern World View and Society, Presented to Martin Stol on the Occasion of his 65th Birthday, 10 November 2005, and His Retirement from The Vrije Universitat Amsterdam*, edited by R. van der Spek and G. Haayer , 235–253. Bethesda: CDL Press.

Horowitz, W. 1989. "An Akkadian Name for Ursa-Minor: mar.gíd.da.an.na = *eriqqi šamāmī*," *Zeitscrift für Assyriologie* 79: 242–5.

______. 2007. "The Astrolabes: Astronomy, Theology, and Chronology." In *Calendars and Years: Astronomy and Time in the Ancient Near East*, edited by J. Steele, 101–13. Oxford: Oxbow Books.

______. 2011. *Mesopotamian Cosmic Geography* (Second Printing, with Corrections and Addenda). Winona Lake, IN: Eisenbrauns.

______. 2014. *The Three Stars Each: The Astrolabes and Related Texts*. Archiv für Orientforschung Beiheft 33. Vienna: University of Vienna.

______. "The Gwich'in Boy in the Moon and Babylonian Astronomy," *Arctic Anthropology* (forthcoming).

Hrůša, I. 2010. *Die akkadische Synonymenliste, malku = šarru : eine Textedition mit Übersetzung und Kommentar*. Alter Orient und Altes Testament (AOAT) 50. Münster: Ugarit-Verlag.

Hunger, H., and T. de Jong. 2014. "The Latest Datable Cuneiform Tablet," *Zeitscrift für Assyriologie* 104: 182–94.

Hunger, H., and D. Pingree. 1989. *MUL.APIN, An Astronomical Compendium in Cuneiform*. Archiv für Orientforschung Beiheft 24. Vienna: University of Vienna.

______. 1999. *Astral Sciences in Mesopotamia*. Leiden: Brill.

Jankoviç, B. 2010. "Estates of Eanna." In *Aspects of the Economic History of Babylonia in the First Millennium BC*, Alter Orient und Altes Testament (AOAT) 377, edited by M. Jursa, 419–28. Münster: Ugarit-Verlag.

Johannes, F., and J. Lemaire. 1999. "Trois tablettes cunéiformes à onomastique ouest-sémitique," *Transeuphratène* 17: 17–34.

Jones, A. 2004. "A Study of Babylonian Observations of Planets Near Normal Stars," *Archive for History of Exact Sciences* 58: 457–536.

Jones, A., and J. M. Steele. 2011. "A New Discovery of a Component of Greek Astrology in Babylonian Tablets: The 'Terms,'" *ISAW Papers* 1.

Kessler, K. 1984. "Eine arsakidenzeitliche Urkunde aus Warka," *Baghdader Mitteilungen* 15: 273–81.

Kidd, D. 1997. *Aratus: Phaenomena*. Cambridge: Cambridge University Press.

Koch, J. 1989. *Neue Untersuchungen zur Topographie des Babylonischen Fixsternhimmels*, Wiesbaden: Otto Harrassowitz.

———. 1999. "Die Planeten-Hypsomata in einem babylonischen Sternenkatalog," *Journal of Near Eastern Studies* 58, 19–31.

Kose, A. 1998. *Uruk. Architektur IV: Von der Seleukiden- bis zur Sasanidenzeit*. Ausgrabungen in Uruk-Warka: Endberichte 17. Mainz: von Zabern.

Krupp, E. C. 1991. *Beyond the Blue Horizon: Myths and Legends of the Sun, Moon, Stars, and Planets*. Oxford: Oxford University Press.

Kurtik, G. E. 2007. *The Star Heaven of Ancient Mesopotamia, the Sumero-Akkadian Names of Constellations and Other Heavenly Bodies*. [In Russian.] St. Petersburg: Aletheia.

Lackenbacher, S. 1977. "Un nouveau fragment de la 'Fête d'Ištar,'" *Revue d'Assyriologie* 71: 39–50.

Lambert, W. G. 1962. "A Catalogue of Texts and Authors," *Journal of Cuneiform Studies* 16: 59–77.

———. 2013. *Babylonian Creation Myths*. Winona Lake, IN: Eisenbrauns .

Linssen, M. 2004. *The Cults of Uruk and Babylon: The Temple Ritual Texts as Evidence for Hellenistic Cult Practice*. Cuneiform Monographs 25. Leiden: Brill/Styx.

Livingstone, A. 1986. *Mystical and Mythological Explanatory Works of Assyrian and Babylonian Scholars*. Oxford: Clarendon Press.

Machinist, P., and H. Tadmor. 1993. "Heavenly Wisdom." In *The Tablet and The Scroll: Near Eastern Studies in Honor of William W. Hallo*, edited by M. E. Cohen, D. C. Snell, and D. B. Weisberg, 146–51. Bethesda: CDL Press.

Meier, G. 1941–1944. "Die zweite Tafel der Serie bīt mēseri," *Archiv für Orientforschung* 14: 139–52.

Monroe, M. W. 2016. "Advice from the Stars: The Micro-zodiac in Seleucid Babylonia." PhD diss., Brown University.

Neugebauer, O. 1947. "Unusual Writings in Seleucid Astronomical Texts," *Journal of Cuneiform Studies* 1: 217–9.

Neugebauer, O., and A. Sachs. 1945. *Mathematical Cuneiform Texts*. New Haven: American Oriental Society.

Nougayrol, J. 1947. "Texts et Documents Figurés," *Revue d'Assyriologie* 41: 23–53.

Ossendrijver, M. 2011. "Science in Action: Networks in Babylonian Astronomy." In *Babylon: Wissenskultur in Orient und Okzident*, edited by E. Cancik-Kirschbaum, M. van Ess, and J. Marzahn, 213–21. Berlin: De Gruyter.

______. "Images of Late Babylonian Cuneiform Signs, Version 6," accessed June 29, 2017, https://www.academia.edu/7302101/2017_Images_of_Late-Babylonian_Cuneiform_Signs_version_6_.

Parpola, S. 1983. *Letters from Assyrian Scholars to the Kings Esarhaddon and Assurbanipal. Part II: Commentary and Appendices*. Kevelaer: Butzon & Bercker.

Pingree, D., and E. Reiner. 1975. "Observational Texts Concerning the Planet Mercury," *Revue d'Assyriologie* 69: 175–80.

Pingree, D., and C. B. F. Walker. 1988. "A Babylonian Star Catalogue: BM 78161." In *A Scientific Humanist: Studies in Memory of Abraham Sachs*, edited by E. Leichty et al., 311–22. Philadelphia: University Museum.

Pongratz-Leisten, B. 1994. *Ina Sulmi Irub, Die kulttopographische und ideologische Programmatik der Akitu-Prozession in Babylonien und Assyrien im I Jahrtausend v. Chr.* Mainz: Philipp von Zabern Verlag.

Reade, J. E. 1986. "Introduction: Rassam's Babylonian Collection: The Excavations and the Archives." In *Catalogue of the Babylonian Tablets in the British Museum. Volume 1: Tablets from Sippar 1*, edited by E. Leichty, xiii–xxxvi. London: British Museum.

Reiner, E. 1995. *Astral Magic in Babylonia*. Transactions of the American Philosophical Society 85/4. Philadelphia: American Philosophical Society.

Reiner, E., and D. Pingree. 1981. *Babylonian Planetary Omens: Part 2*. Malibu: Undena Publications.

Robson, E. 1999. *Mesopotamian Mathematics 2100–1600 BC: Technical Constants in Bureaucracy and Education*. Oxford Editions of Cuneiform Texts 14. Oxford: Oxford University Press.

Rochberg, F. 2004a. *The Heavenly Writing, Divination, Horoscopy, and Astronomy in Mesopotamian Culture*. Cambridge: Cambridge University Press.

______. 2004b. "A Babylonian Rising-Times Scheme in Non-Tabular Astronomical Texts." In *Studies in the History of the Exact Sciences in Honour of David Pingree*, edited by Charles Burnett et al., 56–94. Leiden: Brill.

Rochberg-Halton, F. 1988. "Elements of the Babylonian Contribution to Hellenistic Astronomy," *Journal of the American Oriental Society* 108: 51–62.

Roughton, N. A., and G. L. Canzoneri. 1992. "Babylonian Normal Stars in Sagittarius," *Journal for the History of Astronomy* 23: 193–200.

Roughton, N. A., J. M. Steele, and C. B. F. Walker. 2004. "A Late Babylonian Normal and *Ziqpu* Star Text," *Archive for the History of Exact Sciences* 58: 537–72.

Sachs, A. 1952. "Sirius Dates in Babylonian Astronomical Texts of the Seleucid Period," *Journal of Cuneiform Studies* 6: 105–14.

Sanders, S. 1999. "Writing, Ritual, and Apocalypse: Studies in the Theme of Ascent to Heaven in Ancient Mesopotamia and Second Temple Judaism." Ann Arbor, UMI Dissertation Services.

Schaumberger, J. 1952. "Die *Zipqu*-Gestirne nach neuen Keilschrifttexten," *Zeitscrift für Assyriologie* 50: 214–29.

Steele, J. M. 2006a. "Greek Influence on Babylonian Astronomy?" *Mediterranean Archaeology and Archaeometry* 6: 149–56.

———. 2006b. "Miscellaneous Lunar Tables from Babylon," *Archive for History of Exact Sciences* 60: 123–55.

———. 2007. "Celestial Measurement in Babylonian Astronomy," *Annals of Science* 64: 293–325.

———. 2011. "Astronomy and Culture in Late Babylonian Uruk." In *Archaeoastronomy and Ethnoastronomy: Building Bridges Between Cultures*, edited by C. L. N. Ruggles, 331–41. Cambridge: Cambridge University Press.

———. 2012. "Remarks on the Sources for the Lunar Latitude Section of Atypical Astronomical Cuneiform Text E," *N.A.B.U.* 2012/3, no. 54: 71–2.

———. 2014. "Late Babylonian *Ziqpu*-Star Lists: Written or Remembered Traditions of Knowledge?" In *Traditions of Written Knowledge in Ancient Egypt and Mesopotamia*, Alter Orient und Altes Testament (AOAT) 403, edited by A. Imhausen and D. Bawanypeck, 123–51. Münster: Ugarit Verlag.

———. 2015. "A Late Babylonian Compendium of Calendrical and Stellar Astrology," *Journal of Cuneiform Studies* 67: 187–215.

———. 2017a. *Rising Time Schemes in Babylonian Astronomy*. Dordrecht: Springer.

———. 2017b. "Real and Constructed Time in Babylonian Astral Medicine." In *The Construction of Time in Antiquity: Ritual, Art and Identity*, edited by J. Ben-Dov and L. Doering, 59–82. Cambridge: Cambridge University Press.

Stol, M. 1992. "The Moon as Seen by the Babylonians." In *Natural Phenomena: Their Meaning, Depiction, and Description in the Ancient Near East*, edited by D. J. W. Meijer, 245–77. Amsterdam: Royal Netherlands Academy of Arts and Sciences.

Tallqvist, K. L. 1938. *Akkadische Götterepitheta, mit Götterverzeichnis und einer Liste der prädikativen Elemente der sumerischen Götternamen*. Studia Orientalia 7. Helsinki: Finnish Oriental Society.

van Dijk, J. 1962. "Die Inschriftenfunde." In *Vorläufiger Bericht über die Ausgrabungen in Uruk-Warka 18, Abhandlungen der* Deutschen Orient-Gesellschaft 7, edited by H. J. Lenzen et al., 39–62. Berlin: Gebr. Mann.

Wainer, Z. 2013. "Janus Parallism in Šulgi V," *Bible Lands E-Review* 2013/S2.

Walker, C. B. F. 1995. "The Dalbanna Text: A Mesopotamian Star-List," *Die Welt des Orients* 26: 27–42.

Wallenfels, R. 1993. "Zodiacal Signs among the Seal Impressions from Hellenistic Uruk." In *The Tablet and the Scroll, Near Eastern Studies in Honor of William W. Hallo*, edited by M. E. Cohen, D. C. Snell, and D. B. Weisberg, 281–289. Bethesda: CDL Press.

Watson, R., and W. Horowitz. 2011. *Writing Science before the Greeks: An Naturalistic Analysis of the Babylonian Astronomical Treatise MUL.APIN*. Leiden: Brill.

Weidner, E. 1915. *Handbuch der babylonischen Astronomie*. Leipzig: J.C. Hinrichs'sche Buchandlung.

______. 1925. "Ein astrologischer Kommentar aus Uruk," *Studia Orientalia* 1: 347–58.

______. 1927. "Eine Beschreibung des Sternenhimmels aus Assur," *Archiv für Orientforschung* 4: 73–85.

______. 1963. "Astrologische Geographie im Alten Orient," *Archiv für Orientforschung* 20: 117–21.

______. 1967. *Gestirn-Darstellungen auf Babylonischen Tontafeln*. Graz: Hermann Böhlaus Nachf.

White, G. 2007. *Babylonian Star-Lore, An Illustrated Guide to the Star-lore and Constellations of Ancient Babylonia*. London: Solaria Publications.

Plates

1. Top: VAT 7851 Obv. (copy by E. Weidner reproduced from E. Weidner, *Gestirn-Darstellungen auf babylonischen Tontafeln*, Graz, Hermann Böhlaus, 1967, tafel 2 by permission of the Austrian Academy of Sciences).
Middle: VAT 7847 Obv. (copy by E. Weidner reproduced from E. Weidner, *Gestirn-Darstellungen auf babylonischen Tontafeln*, Graz, Hermann Böhlaus, 1967, tafel 2 by permission of the Austrian Academy of Sciences).
Bottom: AO 6448 Rev. (copy by F. Thureau-Dangin reproduced from F. Thureau-Dangin, *Tablettes d'Uruk*, TCL VI, Paris, Geuthner, 1922).

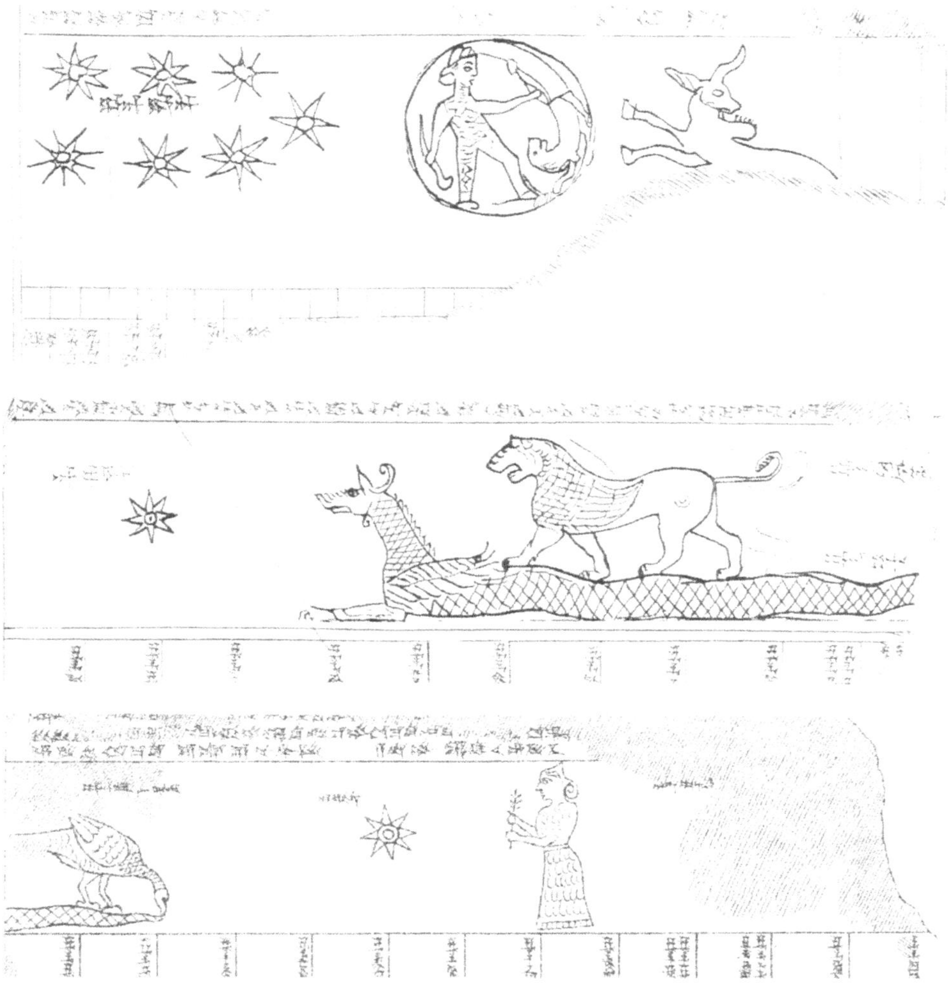

2. VAT 9428 Obv. and lower edge (by permission of the Vorderasiatisches Museum, Berlin).

3. VAT 9428 Rev. and upper edge (by permission of the Vorderasiatisches Museum, Berlin).

4. BM 66958 Side A (top) and Side B (bottom) (© The Trustees of the British Museum).

5. BM 66958 Side A (top) and Side B (bottom) (copy by Wayne Horowitz).

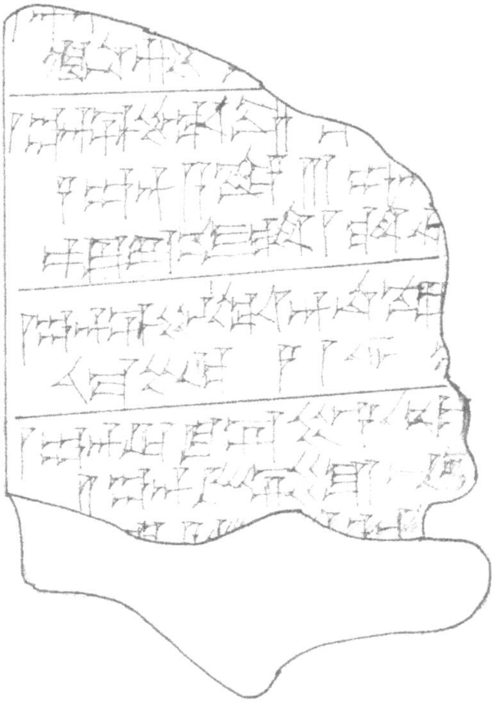

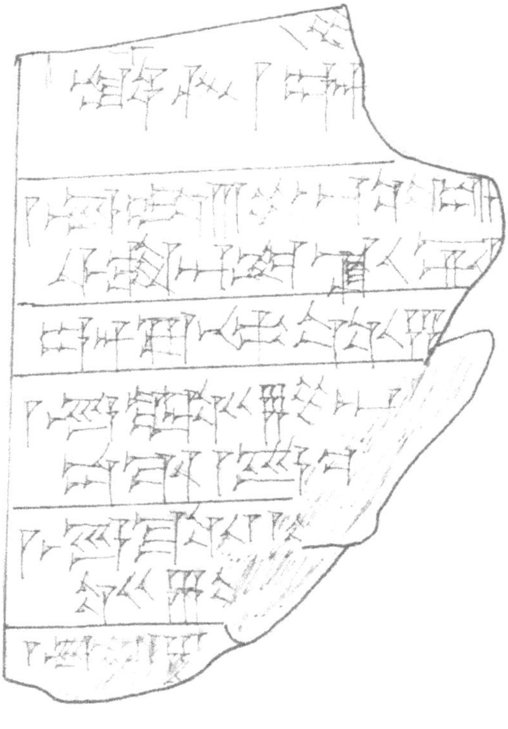

6. NBC 7831 (photograph by Klaus Wagensonner © Yale Babylonian Collection).

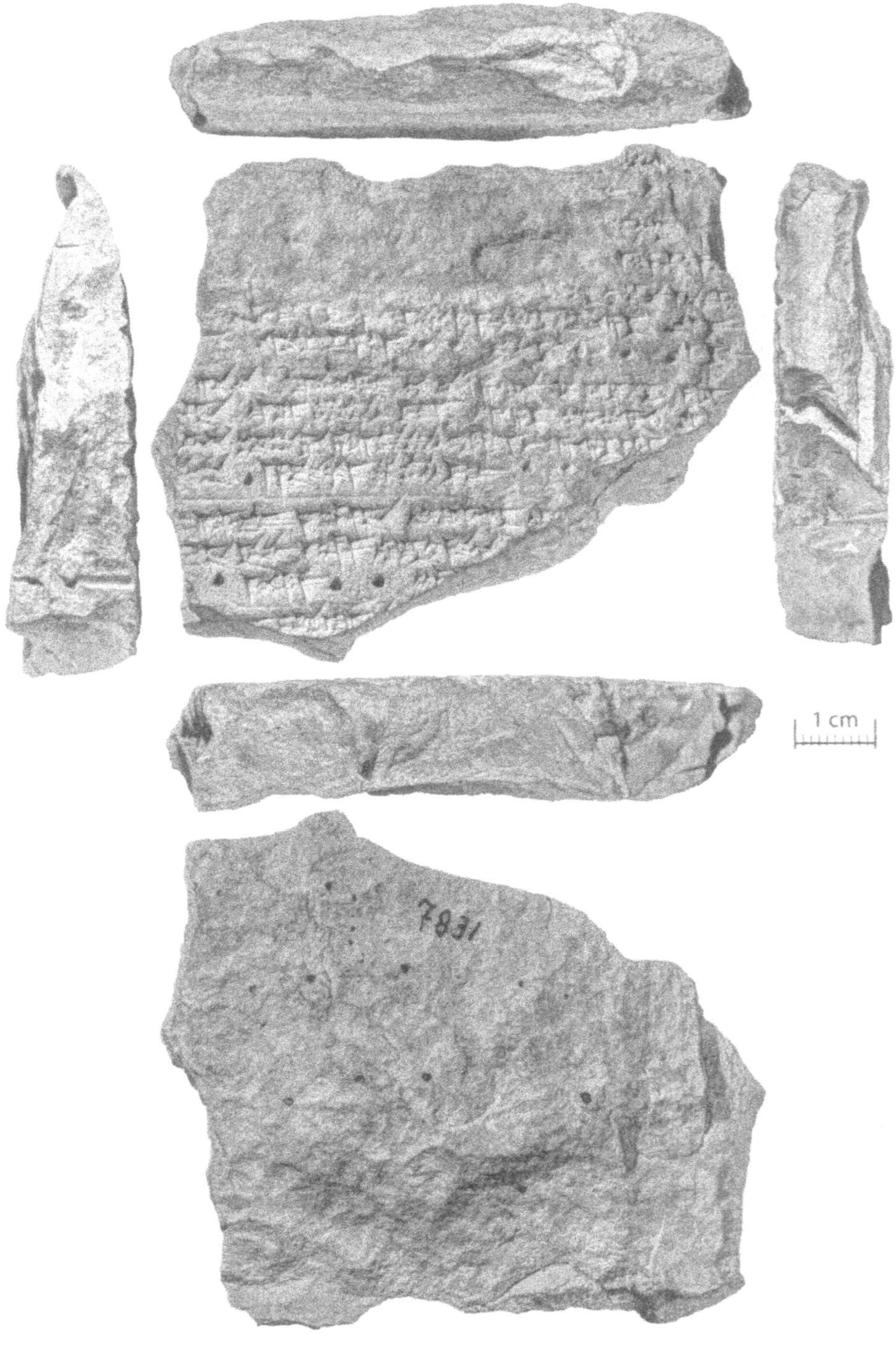

7. NBC 7831 (copy by Paul-Alain Beaulieu).

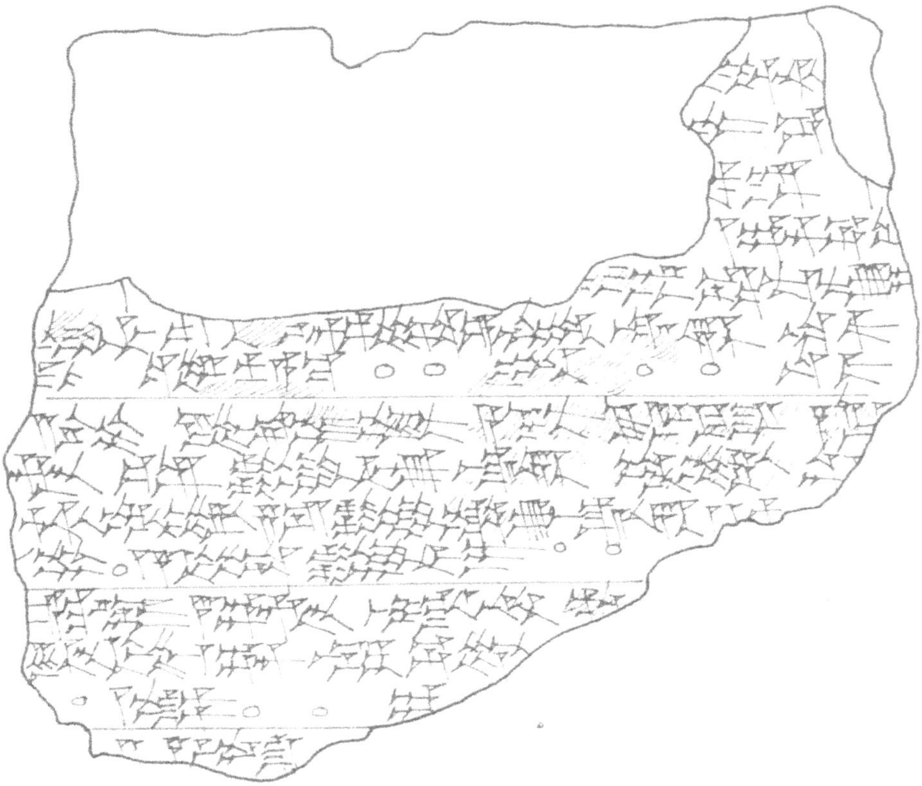

8. MLC 1866 Obv. (photograph by Klaus Wagensonner © Yale Babylonian Collection).

9. MLC 1866 Rev. (photograph by Klaus Wagensonner © Yale Babylonian Collection).

10. MLC 1866 col. i (copy by Paul-Alain Beaulieu).

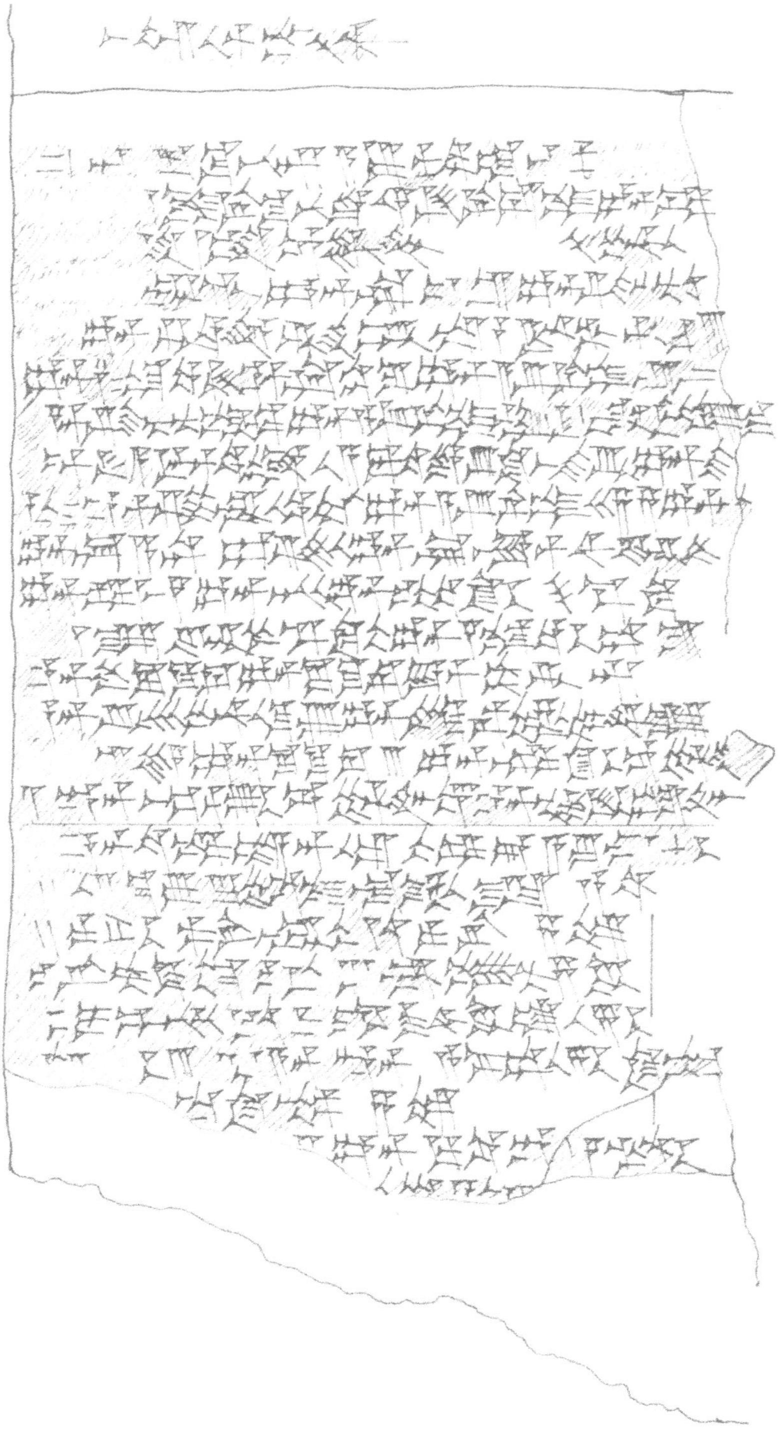

11. MLC 1866 col. ii (copy by Paul-Alain Beaulieu).

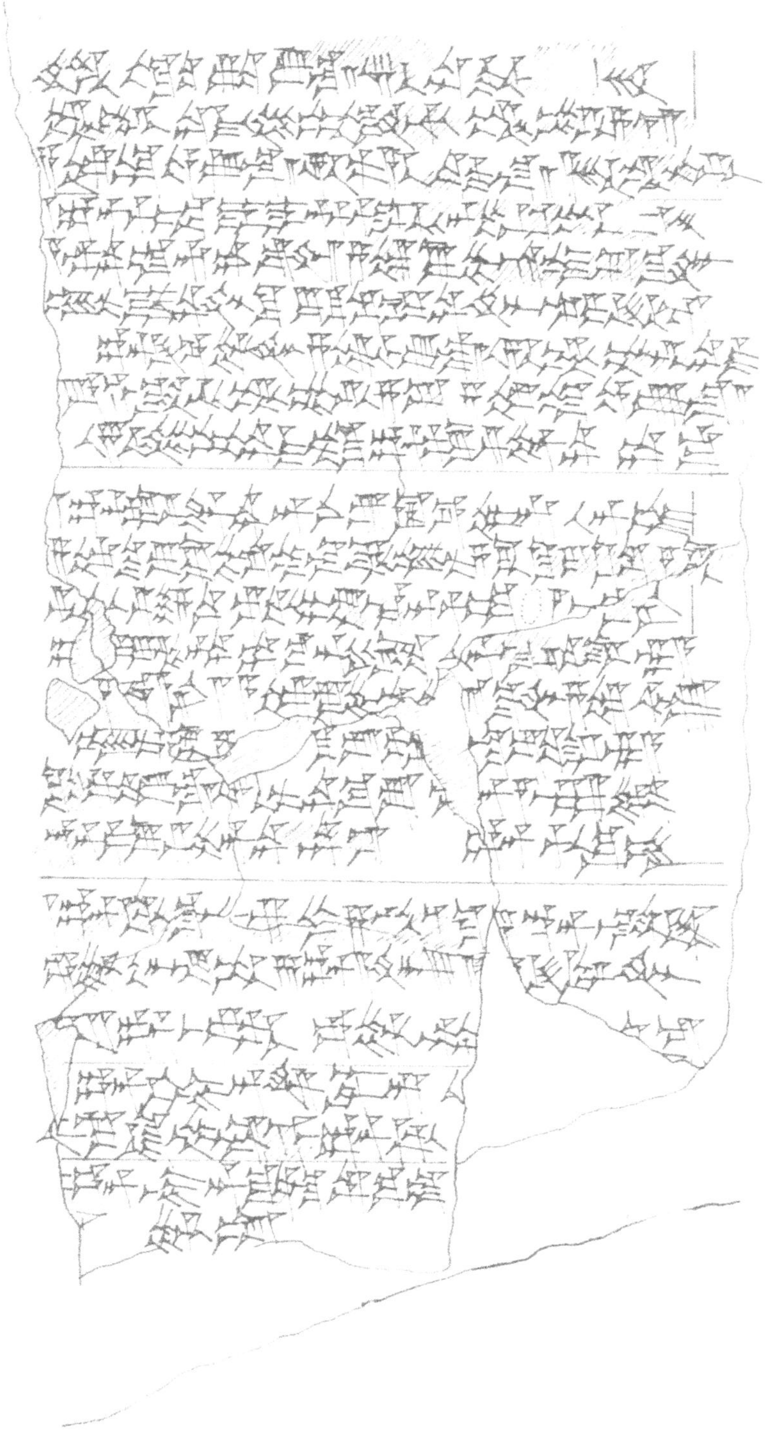

12. **MLC 1866 col. iii (copy by Paul-Alain Beaulieu).**

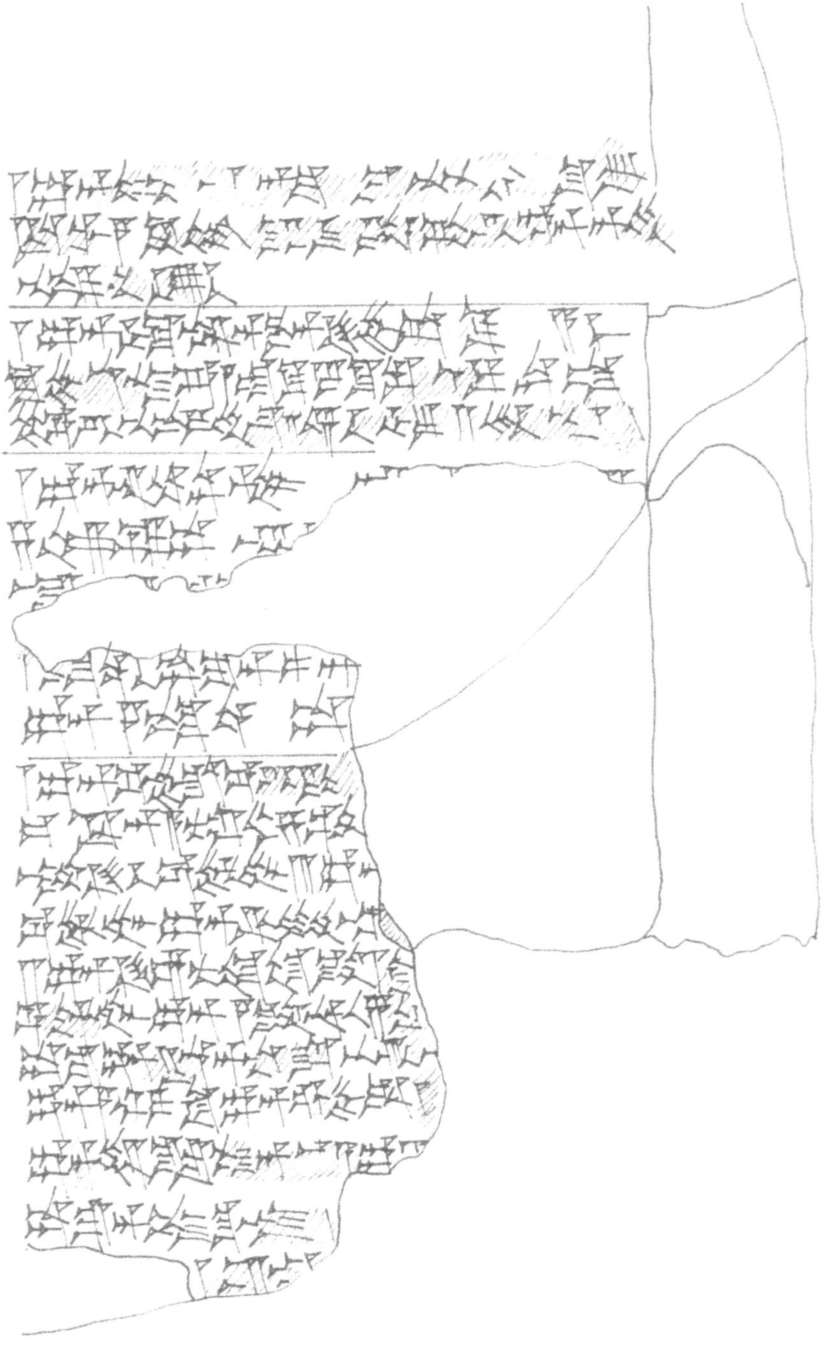

13. MLC 1866 col. iv (copy by Paul-Alain Beaulieu).

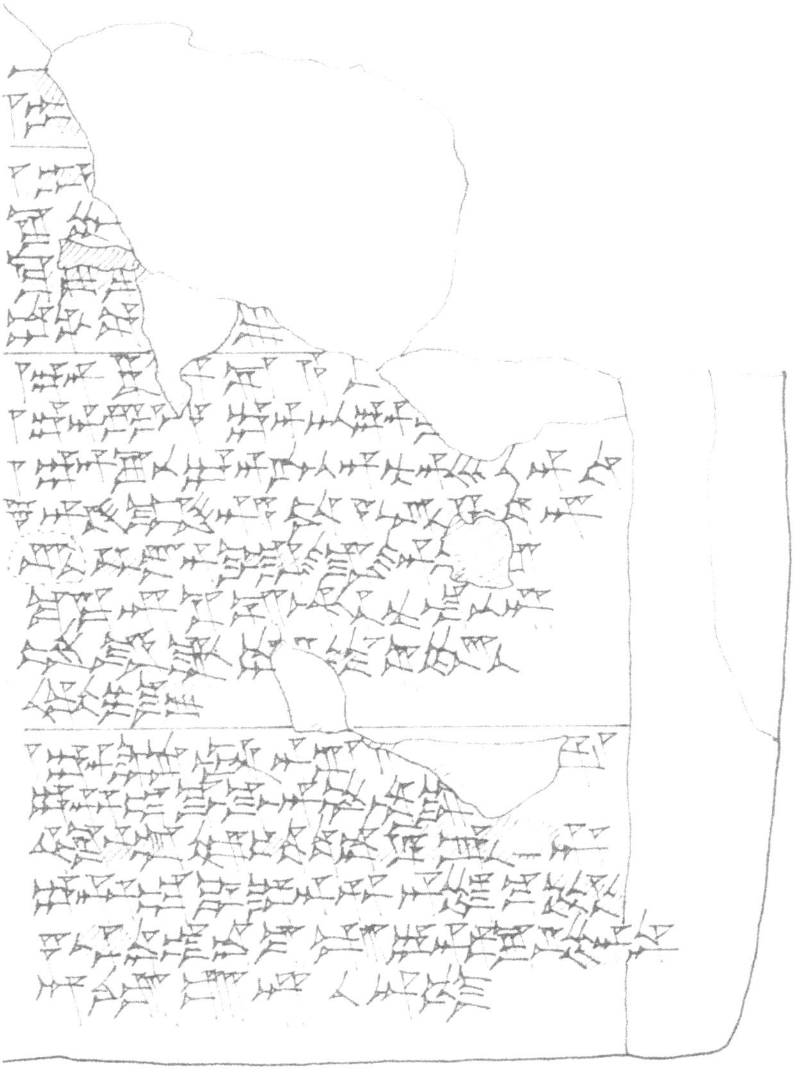

14. MLC 1866 col. v (copy by Paul-Alain Beaulieu).

15. MLC 1884 (photograph by Klaus Wagensonner © Yale Babylonian Collection).

16. MLC 1884 Obv. (copy by Eckart Frahm).

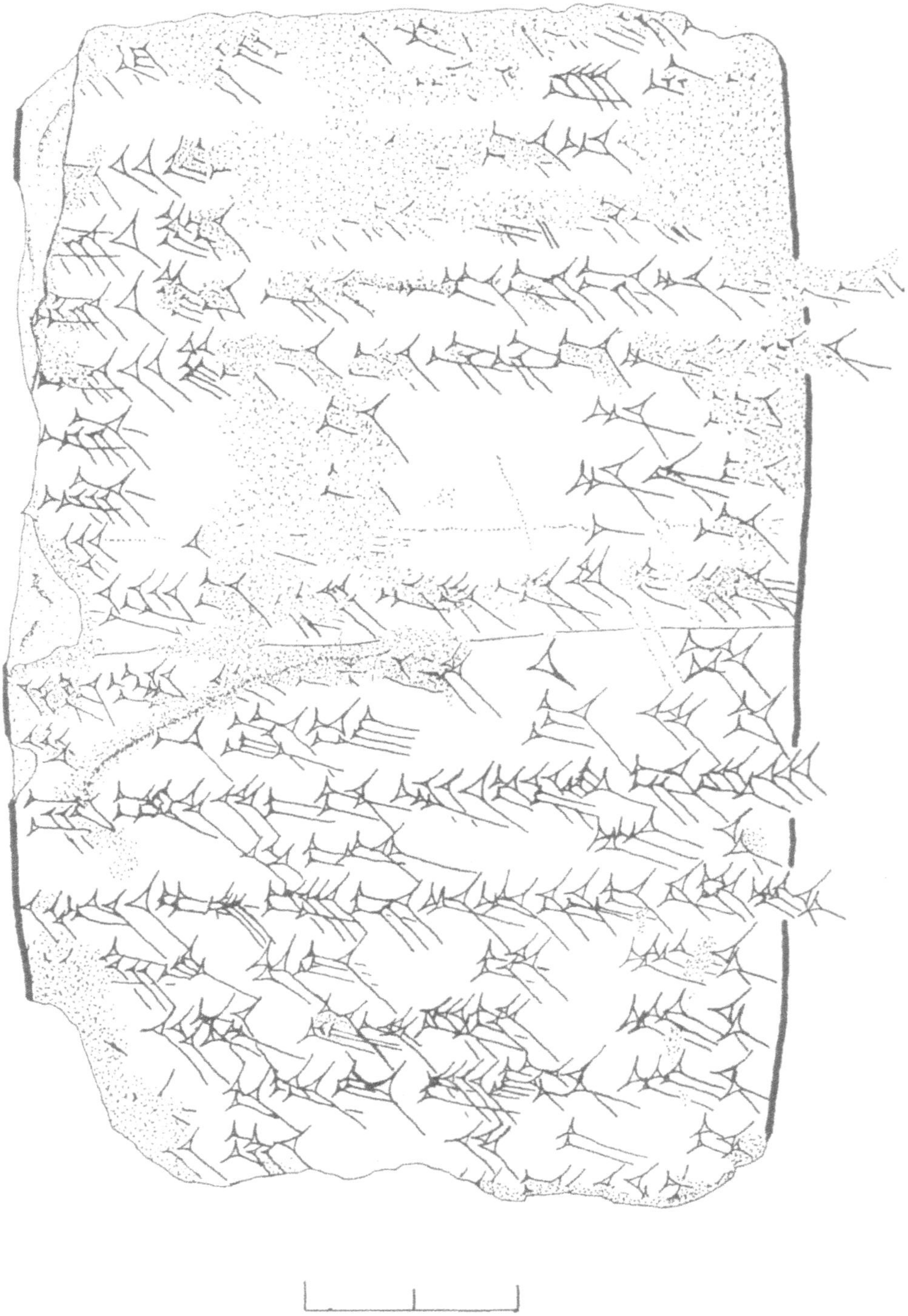

17. MLC 1884 Rev. (copy by Eckart Frahm).

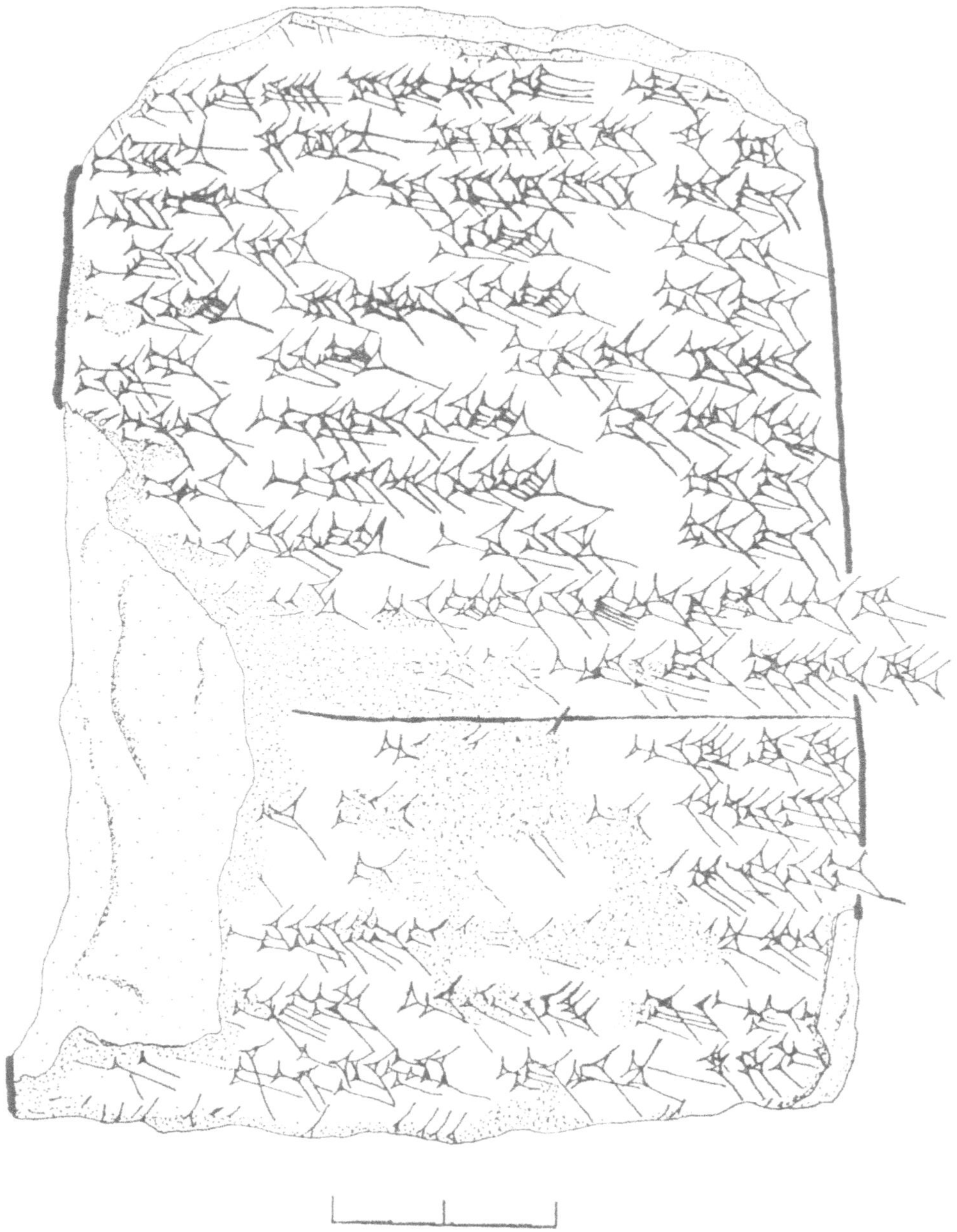

SUBJECT INDEX

INDEX OF STAR-NAMES DISCUSSED[1]

[1] For references to star-names and constellation parts in the main editions of the Uranology Texts see Appendix A: Star Guide (pp. 71–80) with cross referencing to Texts ABCDE and our commentaries.

www.ingramcontent.com/pod-product-compliance
Lightning Source LLC
Chambersburg PA
CBHW061755260326
41914CB00006B/1121
9781606180723